LIVERPOOL JMU LIBRARY

Biology, Rearing, and Care of Young Primates

Biology, Rearing, and Care of Young Primates

James K. Kirkwood and Katherine Stathatos

*Department of Veterinary Science, Institute of Zoology,
Zoological Society of London
London*

Oxford   New York   Tokyo
OXFORD UNIVERSITY PRESS

This book has been printed digitally and produced in a standard specification in order to ensure its continuing availability

OXFORD
UNIVERSITY PRESS

Great Clarendon Street, Oxford OX2 6DP
Oxford University Press is a department of the University of Oxford.
It furthers the University's objective of excellence in research, scholarship,
and education by publishing worldwide in
Oxford New York
Auckland   Cape Town   Dar es Salaam   Hong Kong   Karachi
Kuala Lumpur   Madrid   Melbourne   Mexico City   Nairobi
New Delhi   Shanghai   Taipei   Toronto
With offices in
Argentina Austria Brazil Chile Czech Republic France Greece
Guatemala Hungary Italy Japan South Korea Poland Portugal
Singapore Switzerland Thailand Turkey Ukraine Vietnam

Oxford is a registered trade mark of Oxford University Press
in the UK and in certain other countries

Published in the United States
by Oxford University Press Inc., New York

© J. K. Kirkwood and K. Stathatos 1992

The moral rights of the author have been asserted

Database right Oxford University Press (maker)

Reprinted 2011

All rights reserved. No part of this publication may be reproduced,
stored in a retrieval system, or transmitted, in any form or by any means,
without the prior permission in writing of Oxford University Press,
or as expressly permitted by law, or under terms agreed with the appropriate
reprographics rights organization. Enquiries concerning reproduction
outside the scope of the above should be sent to the Rights Department,
Oxford University Press, at the address above

You must not circulate this book in any other binding or cover
And you must impose this same condition on any acquirer

ISBN 978-0-19-854733-4

Printed and bound in Great Britain by CPI Antony Rowe,
Chippenham and Eastbourne

Preface

Mortality among wild animals bred in captivity is often high during the neonatal and early infant periods. The same is true among wild populations, but we have an opportunity and a responsibility to make efforts to reduce this in animals in our care. Reducing infant mortality can have a significant impact on the rate of population growth and is thus important for the conservation of endangered species represented by only small populations in captivity. Infant mortality rates are related to standards of husbandry and the motive behind this book, which we hope will be the first of a series covering a wide range of orders, was to provide those involved in keeping wild animals with information which will assist them with the management of neonates and young.

The scientific literature relevant to the care of young wild animals is vast, but there have been few attempts to sift through this to assemble the information relevant specifically to management: this has been our aim. When problems arise with captive-born animals it is rarely an opportune time to begin a literature search, and many establishments lack the resources to undertake this in haste. There is a pressing need for practical manuals on wild animal husbandry and there are few yet available. We hope that this contribution will help to fill this gap.

London   J.K.K.
1991   K.S.

Acknowledgements

We are most grateful to the St. Andrew Animal Fund, Sherley's Petcare division of Ashe Consumer Products Ltd., Shamrock Farms (Great Britain) Ltd., Inveresk Research International Ltd., Volac Ltd., and A.T. Twynham and Co for sponsorship. We would also like to thank Jennifer Beard, Dorothy Sherwood, Cedric Chan, Christopher Rigg, and the other students who assisted with literature searches; Mick Carman and other keepers at the Zoological Society of London for discussions on various species; Moya Foreman and Talia Grose for help in the production of the manuscript; Terry Dennett and Mandy Walton for help in preparing the figures and Mr Reg Fish and his assistants Sue Bevis and Alan Baker for their help at the Zoological Society Library.

The Western Tarsier photograph (chapter 5) was supplied by M. Roberts, National Zoological Park, Washington. The Proboscis monkey photograph (chapter 15) is reprinted by permission of the New York Zoological Society. All other photographs were kindly provided by M. Lyster and are reprinted by permission of Zoological Society of London.

We are also grateful to the following for permission to reproduce figures or from whose published data figures were derived: Zoological Society of London (Figs 1.1, 2.1, 6.1, 8.2), American Association of Zoo Veterinarians (Figs 1.2 and 1.3), Academic Press (Figs 3.1, 3.2, 12.1), *Journal of Reproduction and Fertility* (Fig. 4.1), S Karger AG, Basel (Figs 5.1, 7.5, 10.2, 13.2), Laboratory Animals (Figs 6.2, 13.1, 13.3), Primates (Figs 7.1, 14.2), American Association for Laboratory Animal Science (Figs 7.2, 10.2), Longman (Fig. 9.1), John Wiley and Sons (Figs 13.1, 18.2), Zoologischer Garten Koln (Figs 7.2, 7.3), Plenum Publishing Corporation (Figs 14.1, 17.2), Jersey Wildlife Preservation Trust (Fig. 16.2), the Carnegie Institution (Fig. 17.1), Dr M. Perret (Fig. 2.1), Prof G.A. Doyle (Figs 3.1, 3.2), Dr Laura Dronzek (Figs 7.2, 7.3), Herr W. Kaumanns (Figs 7.2, 7.3), Dr Lynne Ausman (Fig. 10.2), Dr G.T. Moore (Fig. 14.1), Dr G.C. Ruppenthal (Fig. 12.3). Figure 8.1 was redrawn by permission of the Smithsonian Institution Press from *Smithsonian Contributions to Zoology*, **354**, by R.J. Hoage. © Smithsonian Institution, Washington, D.C. 1982, p. 41, figure 23.

Contents

	Introduction	1
1	Ruffed lemur	7
2	Lesser mouse lemur	15
3	Senegal or lesser bush-baby	21
4	Bosman's potto	27
5	Horsfield's or western tarsier	33
6	Common marmoset	39
7	Cotton-top tamarin	49
8	Golden lion tamarin	59
9	Owl monkey (night monkey or douroucouli)	65
10	Squirrel monkey	71
11	Vervet monkey (Green monkey or grivet)	81
12	Rhesus macaque	89
13	Stump-tailed macaque	99
14	Common baboon	107
15	Proboscis monkey	119
16	Western black and white colobus monkey	125
17	Lar gibbon	133
18	Chimpanzee	143
	List of products	151
	Index	153

Introduction

Primates in captivity

Captive propagation is an important component of primate conservation (Mittermeier 1982; Foose *et al.* 1987). Although keeping primates for display in zoos, for research, and as pets has a long history, it is only relatively recently that species have been bred in captivity on a large scale. Progress in captive primate husbandry, driven by the need for healthy animals for research and, more recently, by the recognition of the importance of captive breeding as part of a conservation strategy, has been remarkably rapid. It is surprising, now that marmosets are bred in large numbers in research and zoo colonies all over the world, that just 25 years ago Lee S. Crandall (1964) observed in his book on the management of wild mammals in captivity that: 'General and usually unhappy experience has convinced most zoo authorities that marmosets are too delicate to survive under ordinary conditions.'

It is difficult to estimate the total number of primates maintained in captivity but it is probably at least 20 000 potential breeding animals (Schmidt 1986) and quite possibly far more. About three-quarters of the some 184 primate species (Martin 1990) have bred in captivity (Jones 1986) but a much smaller proportion are kept in sufficient numbers to ensure maintenance of genetic diversity for the distant future (Flesness 1986; Schmidt 1986; Foose *et al.* 1987). The largest captive populations are of species widely used for biomedical research.

Neonatal and infant mortality

For both conservation and welfare reasons it is important to strive for improvements in primate husbandry. Analysis of colony records has frequently revealed a high level of mortality in the neonatal and early juvenile periods, as the subsequent species by species accounts reveal (see also Flesness 1986; Lindburg *et al.* 1986). In some species, neonatal mortality is a major factor limiting population growth. For example, from an analysis of the genetic and demographic status of the western lowland gorilla in captivity, Mace (1988) demonstrated that there could be a gradual decline in the population. However, reducing first-year mortality to less than 20 per cent would help to ensure a stable population or slowly increasing numbers.

There are many causes of infant mortality and identification of these requires both post-mortem examination of individuals and assessment of management. In many cases, suboptimal management is the root cause. Previous social and breeding history often has a critical influence on the maternal behaviour in primates and thus on their ability to rear young. In turn, previous social history is dependent on the system of management.

Husbandry of young primates: the reason for this book

There is already a vast literature on primate management and it is growing rapidly. One of the difficulties of those involved in primate breeding is tracking down and obtaining the available information. The motive behind writing this book was to try to assist in reducing infant mortality in captive primates by assembling information relevant to the successful rearing of parentally competent animals. This has not been an easy task because of the volume of the literature and because authorities not infrequently differ in opinion about techniques. For some species, a great deal of information is available whilst for others, there is very little. Another difficulty has been in deciding the level at which to pitch the information: some primates are bred in academically oriented specialist centres, some in zoos, and some in small private collections, and some as pets.

Our approach has been to review information relevant to the captive management of a representative range of species for which data are available. Inevitably, coverage of specific aspects, such as infant nutrition and techniques for re-introduction into family groups, is uneven and we have to leave to the reader's judgement the extent to which techniques described for one species are appropriate for another. As a first approximation, management techniques for one species are likely to be appropriate for others that are closely related.

We have begun each section with notes on the status and distribution of each species. This is not relevant to the technology of rearing infants but puts efforts in this direction into perspective. We have also provided short notes on basic information about sex ratio, gestation period, breeding season, breeding age, longevity, and other aspects that provide a useful background in assessing causes of infant mortality. The main emphasis has, however, been placed in providing details on infant management, such as nutrition and feeding techniques, accommodation and reintegration of artificially reared infants into peer or family groups. We have also assembled data, where available, on growth and development, and on the incidence and causes of disease and mortality.

Artificial rearing is an important aspect of captive primate husbandry, enabling abandoned or orphaned infants that would otherwise be lost, to be saved in order to contribute to the breeding population. However, because of the importance of social interaction in the normal development of primates, artificial rearing presents particular difficulties (although there are species differences in the sensitivity of subsequent breeding and maternal behaviour to disturbance by the environment during rearing, and prosimians appear to be more robust than the others in this respect). Techniques have been devised for artificially rearing reproductively competent animals of a wide range of species.

Hand-rearing primates remains, however, a difficult, labour-intensive and, because of their slow growth rate, protracted procedure. Unless the facilities and resources are available to meet the physiological and behavioural needs of the infant, euthanasia should be considered as perhaps a more humane alternative. Another option that is sometimes available is cross-fostering and techniques for this have also been mentioned where data are available.

Growth rates of primates

Although the prosimians have growth rates comparable with many other mammals of similar adult size, the Anthropoidea tend to grow more slowly. The Old World monkeys and the apes have the slowest growth rates (Kirkwood 1985). We have drawn growth curves for each species from data published in the literature, which provide a yardstick against which the progress of individuals can be judged.

The variation in growth rates between species of primates is associated with differences in milk composition (those with higher growth rates, notably the prosimians, tend to have higher protein concentrations in the milk), and with energy intake rate during growth.

Energy intake during growth

When artificially rearing wild animals, it is important to have an estimate of their food requirement. Judging the degree of hunger and satiation in very young mammals is not easy and there is some evidence that appetite and optimum requirement do not necessarily correspond. The milk intake of mother-reared infants is limited and controlled partly by the mother. Underfeeding slows or stops growth and leads to debility and increased susceptibility to infectious diseases and opportunist pathogens. Overfeeding may cause digestive disturbances which can have severe consequences. These are common problems in hand-reared primates and therefore we have placed emphasis on providing information on milk and energy intakes.

The daily energy requirements of growing animals can be visualized as having two components: the amount necessary to maintain the animal at its present size, and the amount needed for growth. In rapidly growing animals the component for growth often exceeds that required for maintenance and the total energy intake rate is high in relation to the size of the animal. In the slow growing primates the amount of energy required for growth each day tends to be very low, and the total daily energy intake may exceed the maintenance requirement by very little (Kirkwood 1985). Methods for estimating total requirement have been described by Kirkwood (1991) Kirkwood and Mace (in press), and Kirkwood and Bennett (1992).

The energy requirements of animals do not increase in direct proportion to body weight, but increase by about 68 per cent with each doubling in weight between species. This can be expressed in a different way: energy requirements increase with body weight raised to the 0.75 power. Kleiber (1975) proposed that energy requirements should therefore be calculated in relation to 'metabolic weight' (weight in kilograms, raised to the 0.75 power). We share the view that this is a concept useful for predicting the energy and thus food requirements of species for which no such data are available. We have therefore often expressed energy intakes in relation to metabolic weight. These figures may prove useful guides for predicting the requirements of closely related species. These predictions can be made quite simply using an electronic calculator with exponent function and do not require that the concept of 'metabolic weight' is fully understood. Estimates of energy requirements made in this way should be treated with caution as being first approximations.

Application for other species

We have included 18 species in this book, including representatives of nine out of the 11 primate families. Their taxonomic positions are indicated in the table below. The families not represented are the Indriidae and the Daubentoniidae. These comprise a total of five species, all of which are rare in captivity and on which there is as yet very little published information on management.

As we have mentioned above, the degree to

The taxonomic position of species

	No. of species	Representatives
Order Primates		
Suborder Prosimii		
Family Lemuridae	10	Ruffed lemur
Cheirogalidae	7	Lesser mouse lemur
Indriidae	4	
Daubentonidae	1	
Lorisidae	10	Senegal or lesser bush-baby
		Bosman's Potto
Tarsiidae	3	Horsefield's or western tarsier
Suborder Anthropoidea		
Family Callitrichidae	21	Common marmoset
		Cotton-top tamarin
		Golden lion tamarin
Cebidae	30	Owl monkey (night monkey or douroucouli)
		Squirrel monkey
Cercopithecidae	82	Vervet or grivet monkey
		Stump-tailed macaque
		Rhesus macaque
		Common baboon
		Proboscis monkey
		Western black and white colobus monkey
Pongidae	4	Chimpanzee
Hylobatidae	9	Lar gibbon

which management practices can be extrapolated between species (for example, those found suitable for the squirrel monkey and owl monkey to the other 28 members of the family Cebidae) we have to leave to the reader's judgement. Many of the most important aspects of management apply widely. Characteristics relevant to management which do differ between species include: milk composition, daily milk requirements, growth rates, and patterns of behavioural development. Techniques of management to provide socialization during growth also vary between species. Although specific information on the rearing management of most of the species not included in this book is sparse, there are some species for which there is a considerable amount of literature. Major sources of information are referred to in the relevant sections.

References

Crandall, L.S. (1964). *The management of wild mammals in captivity*. University of Chicago Press.

Flesness, N.R. (1986). Captive status and genetic considerations. In *Primates. The road to self-sustaining populations* (ed. K. Benirschke), pp. 845–56. Springer-Verlag, New York.

Foose, T.J., Seal, U.S., and Flesness, N.R. (1987). Captive propagation as a component of conservation strategies for endangered primates. In *Primate conservation in the tropical rain forest* (C.W. ed. Marsh, and R.A. Mittermeier), pp. 263–99. Alan R. Liss, New York.

Jones, M.L. (1986). Successes and failures of captive breeding. In *Primates. The road to self-sustaining populations* (ed. K. Benirschke), pp. 251–60. Springer-Verlag, New York.

Kirkwood, J.K. (1985). Patterns of growth in primates. *Journal of Zoology, London*, **205**, 123–36.

Kirkwood, J.K. (1991) Energy requirements for maintenance and growth of wild mammals, birds and reptiles in captivity. *Journal of Nutrition* **121**, S29–S34.

Kirkwood, J.K. and Bennett, P.M. (1992) Approaches and limitations to the prediction of energy requirements in wild animal husbandry and veterinary care. *Proceedings of the Nutrition Society* (in Press).

Kirkwood, J.K. and Mace, G.M. (in press). Patterns of growth in mammals. In *The management of wild mammals in captivity*, Vol. 1 (ed. D.G. Kleiman). Sinauer, Massachusetts.

Kleiber, M. (1975). *The fire of life*. Krieger, New York.

Lindburg, D.G., Berkson, J., and Nighthelser, L. (1986). The contribution of zoos to primate conservation. In *Primate ecology and conservation*, Vol. 2 (ed. J.G. Else and P.C. Lee), pp. 295–300. Cambridge University Press.

Mace, G.M. (1988). The genetic and demographic status of the Western lowland gorilla (*Gorilla g. gorilla*) in captivity. *Journal of Zoology, London*, **216**, 629–54.

Martin, R.D. (1990). *Primate origins and evolution*. Chapman & Hall, London.

Schmidt, C.R. (1986). A review of zoo breeding programmes for primates. *International Zoo Yearbook* **24/25**, 107–23.

Seal, U.S. and Makey, D.G. (1974) *ISIS Mammalian taxonomic directory*. Minnesota Zoological Garden, St Paul.

Ruffed lemur

1 Ruffed lemur

Species
The ruffed lemur   *Varecia variegata*

ISIS No.   1406001002005001

Status, subspecies, and distribution
There are two subspecies *Varecia variegata variegata*, the black and white ruffed lemur and *V.v. rubra*, the red ruffed lemur. The former occurs in eastern madagascar and the latter in a restricted area in the north-east of the island (Jolly 1986). The species is classified as endangered International Union for the Conservation of Nature, (IUCN 1990). Numbers in the wild have declined rapidly Tattersall 1982).

The world captive population of ruffed lemurs in 1985 was estimated at 462 by Pollock (1986), and Brockman (1986) reported 358 *V.v. variegata* and 125 *V.v. rubra* registered in the *International studbook*.

Sex ratio
Of 481 births of *V.v. variegata* recorded in the *International studbook* (Brockman 1986), 266 were males and 215 were females. This suggests a sex ratio at birth of 1.2 males to 1 female. The records in this studbook for *V.v. rubra* show 85 male and 79 female births.

Social structure
In the wild, ruffed lemurs live in family groups and juveniles leave their parents after their first year. Groups of more than three individuals have rarely been reported which suggests that, in view of the fact that multiple births occur, infant mortality may be high (Tattersall 1982).

Breeding age
Some males may mate with and impregnate females in their second year (Boskoff 1977; Tattersall 1982). However, males are considered fully mature at about 4 years of age when their testes reach maximum size during the breeding season (Foerg 1982). Females usually begin oestrous cycles at 18 to 22 months of age (Boskoff 1977; Tattersall 1982; Brockman *et al.* 1987*b*) and are thus able to breed in the first breeding season of their second year (Shideler and Lindburg 1982), although such early breeding may not always be successful (Foerg 1982).

Longevity
There are records of both black and white and red ruffed lemurs living to 26 years of age (Brockman 1986).

Seasonality
In Madagascar, births occur in October and November (Tattersall 1982). In captive animals in the northern hemisphere, the mating season extends from mid-December through to April and most births occur in April and May (Shideler and Lindburg 1982; Brockman *et al.* 1987*b*). The female ruffed lemur is seasonally polyoestrus and shows up to three cycles with spontaneous ovulation if conception does not occur (Shideler and Lindburg 1982; Brockman *et al.* 1987*b*). In the colony at the Zoological Society of San Diego, 64 per cent of females conceived at the first oestrus of the season (Brockman *et al.* 1987*b*). There are visible changes in the morphology of the vulva during the mating season (Foerg 1982); monitoring these can enable determination of the day of ovulation, conception, and prediction of parturition date (Brockman *et al.* 1987*b*). As oestrus approaches, the vulva, which during anoestrus is black and imperforate, begins to form a pink oval opening. Mating is limited to a one day period at the time when the vulva size reaches a peak (Foerg 1982; Brockman *et al.* 1987*b*). The three oestrous periods occur at 30 to 40 day intervals (Bogart *et al.* 1977*a, b*).

Males show seasonal changes in testicle size.

The testes begin to increase in size about two months prior to oestrus in the females (Bogart et al. 1977a).

Gestation

Gestation varies between 90 and 110 days (Hick 1976; Boskoff 1977; Lindsay 1977; Shideler and Lindburg 1982), with a mean of 102 days (Foerg 1982; Brockman et al. 1987b).

Pregnancy diagnosis

Females may exhibit unusual aggression in early pregnancy, and abdominal distension becomes visible at about 80 days of gestation (Lindsay 1977). Fetuses can be palpated at 60 days (Boskoff 1977), or quite possibly at an earlier stage. The nipples may become engorged, erect, and thus visible at 6 to 8 weeks of gestation (Brockman et al. 1987b), but Shideler and Lindburg (1982) indicated that this only becomes noticeable after birth when the mothers are nursing.

Birth

Boskoff (1977) and Lindsay (1977) reported births during the daylight hours, but at San Diego most births have occurred between 20.00 h and 04.00 h (Meier and Willis 1984; Brockman et al. 1987b). The young are delivered and deposited in nests. At San Diego most babies were born in the nest boxes provided but some were born elsewhere in the cage where they were at greater risk of hypothermia (Brockman et al. 1987b). The mother eats the placenta and then cleans the young before manipulating them under her (Meier and Willis 1984; Brockman et al. 1987b). The process of labour has been described by Foerg (1982). The female sat on her hind legs, strained with contractions, and licked her genitals. Half an hour after contractions were first observed the tail or a limb of the first baby appeared. The mother cleaned the first baby before the second was born 10 minutes later. Some females squeaked and whimpered during parturition.

Litter size

Litter size varies from 1 to 5 but is usually less than 4 (Boskoff 1977; Tattersall 1982; Brockman et al. 1987b). At San Diego the most common litter size was triplets (40.5 per cent) but in the captive population as a whole twins were found to be most frequent (36.5 per cent), followed by singles (28.2 per cent) (Brockman et al. 1987b). The latter authors suggested that inbreeding depression may be involved in variation in litter size between colonies, but nutrition could play a part as is likely, for example, in cotton-top tamarins. Foerg (1982) found that parity had a significant influence on litter size; the first pregnancy of 7 out of 10 females resulted in singleton births.

Adult weight

Ruffed lemurs are the largest of the lemur family, and adult weight ranges from 3.2 to 4.8 kg (Bogart et al. 1977a; Boskoff 1977; Tattersall 1982).

Neonate weight

Birth weights range from 80 to 125 g in viable infants. Babies weighing less than 70 to 80 g rarely survive without intervention. At San Diego, the mean birth weight of *V.v. rubra* was 95.7 g ($n = 21$) in 1983, but increased to 101.7 g ($n = 15$) in 1984. *Varecia v. variegata* also showed an increase in mean neonate weight from 83 g ($n = 13$) in 1983 to 92.7 g ($n = 10$) in 1984 (Brockman et al. 1987b). Although we are not aware of any data on the subject, it would seem likely that neonate weight is inversely correlated with litter size.

Adult diet

In the wild, ruffed lemurs are primarily frugivorous although they probably also eat leaves, flowers, and other vegetable matter, according to availability (Tattersall 1982).

In captivity, animals have been fed on commercial primate pellets supplemented with a variety of fresh plant foods such as leaves and branches (*Hibiscus, Eugenia,* and *Syzygium*), vegetables (spinach, cabbage, celery, cooked green beans, sweet potatoes, and carrots), and fruit (Lindsay 1977; Brockman et al. 1987a).

Adult energy requirements

Pollock (1986) reviewed the literature on basal metabolic rates and voluntary food intake in prosimians and pointed out that in all species studied the basal metabolic rate was lower than

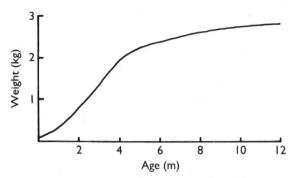

Fig. 1.1. Mean growth rate of twin hand-reared ruffed lemurs to 365 days. From Cartmill *et al.* (1979).

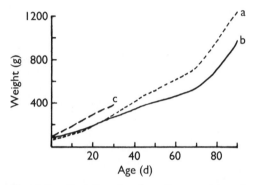

Fig. 1.2. Growth rate of mother- and hand-reared ruffed lemurs to 90 days. From Meier and Willis (1984). (a) *Varecia v. variegata* hand-reared, (b) *V.v. rubra* hand-reared. (c) *V.v. variegata* mother-reared.

the 'Kleiber mean' of 70 kcal/d per kg$^{0.75}$ and his data base showed that voluntary food intakes were correspondingly low. As far as we are aware there have been no measurements of energy metabolism in *V. variegata* but, in view of the above, it is possible that daily energy intake for maintenance may be as little as 100 kcal/d kg per$^{0.75}$ that is, about 280 kcal/d for a 4 kg individual.

Growth

Patterns of weight gain of some hand-reared ruffed lemurs have been described by Cartmill *et al.* (1977) and Meier and Willis (1984). These animals reached 1 kg at 70 to 90 days of age and grew at about 15 g/d between 1 and 4 months of age. Growth rate slowed thereafter (Figs 1.1 and 1.2).

The lemurs hand-reared by Meier and Willis (1984) did not grow as fast during the first month as those reared by their mothers (Fig. 1.2), but achieved similar adult weights. The initially slower growth rates may have been due to sequels of the neonatal problems that led to their being taken for hand-rearing, and perhaps because initial feeding frequency (every 2 hours) was suboptimal as babies are nursed more or less continuously by their mothers during the first 2 weeks (Meier and Willis 1984). The composition of the milk replacer used may not have been ideal, although these authors reported that on this diet, the survival of hand-reared babies that weighed more than 70 g at birth was excellent (no statistics were provided).

Four reports of hand-rearing *Varecia variegata* were included in the AAZPA *Infant diet/care notebook* (Taylor and Bietz 1985). The growth of three of the four babies described was similar to that shown in Fig. 1.2, but the fourth grew considerably faster reaching 701 g at 50 days of age.

Milk and milk intake

As far as we are aware no data on the milk composition of ruffed lemurs are available. However, the composition of milk samples from *Lemur catta*, *L. fulvus*, and *L. macaco* has been reported by Buss *et al.* (1976). The mean composition of these milks was dry matter 11.7 per cent, fat 2.3 per cent, protein 2.7 per cent, lactose 6.4 per cent and ash 0.35 per cent. The proportions of fat, protein, lactose, and ash in the dry matter were thus 0.20, 0.23, 0.54, and 0.03 respectively. The energy content of this mixture of lemur milks was estimated to be 0.56 kcal/ml, which is lower than that of most primates. As a guide to the composition of the milk of *Varecia* these data should be treated with some caution. *Varecia* may differ because its litter sizes tend to be larger and also because it has a different infant care strategy, 'parking' its litter in a nest instead of carrying infants constantly (Cartmill *et al.* 1979).

A variety of milk substitutes have been used to rear ruffed lemurs. Meier and Willis (1984) fed a formula consisting of 120 ml whole cow's milk plus 15 ml rice cereal (to which 1–2 drops of a paediatric vitamin preparation were added once daily) for the first 2 weeks and thereafter, 120 ml whole cow's milk plus 22.5 ml rice cereal. The composition of this mixture was 3.2 per cent fat, 2.9 per cent protein, and 6.6 per cent carbohydrate. Cartmill et al. (1979) used a replacer made from tinned condensed milk, whole cow's milk, water, and Meritene (a soluble invalid nutrient powder) mixed in the ratio 12:12:3:1. (They noted that constipation occurred when skimmed milk was substituted for the whole milk.) Brockman et al. (1987b) described a formula essentially the same as that used by Meier and Willis. Formulae based on cow's milk with added rice cereal and, in one case egg yolk, have been reported in the AAZPA *Infant diet/care notebook* (Taylor and Bietz 1985). The fat, protein, and carbohydrate proportions of the lemur milk reported above are close to those in Primilac, so this may prove a suitable off-the-shelf milk replacer for the larger lemurs.

The literature indicates formula intake rising from 12 to 40 ml on the first day (means of 24 and 32 ml were noted by Meier and Willis 1984, for *V.v. rubra* and *V.v. variegata* respectively), to means of 150 to 200 ml/d by 3 months of age (see Fig. 1.3).

Daily milk energy intake in the lemurs reared by Meier and Willis (1984) rose to a mean of about 250 kcal/kg$^{0.75}$ at about 10 days of age, remained at this level until 30 to 40 days of age, then gradually declined to 130 kcal/kg$^{0.75}$ at 90 days of age (see Fig. 1.4). The rate of decline will obviously depend upon the quantity of solid food eaten as weaning progresses.

Lactation and weaning

Unlike other lemur species, *Varecia* possess three pairs of mammary glands and lactate from all of them (Boskoff 1977; Tattersall 1982). Newborn babies have been observed to start sucking within 30 minutes of birth (Foerg 1982). Mother-reared babies were first seen to take solid foods at 25 days of age (Lindsay 1977). At the San Diego colony, infants are allowed to continue sucking for 6 months but they are then weaned by taking them away from their mothers for 14 days or until lactation ceased (Brockman et al. 1987b). The rationale for this was to give the mother the best

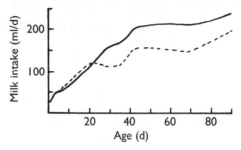

Fig. 1.3. Approximate daily milk intake of hand-reared ruffed lemurs in relation to age: solid line, *Varecia v. rubra*; broken line, *Varecia v. variegata*. From Meier and Willis (1984).

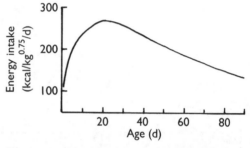

Fig. 1.4. Approximate daily energy intake provided by milk in hand-reared ruffed lemurs expressed per unit of metabolic weight, in relation to age (pooled data for *Varecia v. rubra* and *Varecia v. variegata*). From Meier and Willis (1984).

opportunity to regain good body condition for the next breeding season.

Hand-reared babies have been offered solid foods starting at about 3 weeks of age, the age at which mother-reared babies start sampling solid foods (Meier and Willis 1984). Cartmill et al. (1979) introduced puréed fruits and vegetables on day 18, baby cereal and chopped fruits on day 24, and commercial primate chow mixed with banana and fruit juice on day 74. At San Diego, a mush of soaked primate pellets mixed with banana and fruit juice was first provided at 3 to 4 weeks of age (Meier and Willis 1984; Brockman et al. 1979b). Solid food intake increased rapidly at 6 weeks of age (Meier and Willis 1984). Weaning has usually been completed at 4.5 to 5 months of age (Cartmill et al. 1979; Brockman et al. 1987b), but there are accounts of weaning at 3 months (Taylor and Bietz 1985).

Feeding

For the first two or more feeds, or until a good feeding response has been established, babies taken for hand-rearing have been given 5 per cent glucose solution (Meier and Willis, 1984; Brockman et al. 1987b). These authors adopted a regime of 1 to 2 hourly feeds throughout the day and night until solids were beginning to contribute significantly to the daily requirements, or when a body weight of about 500 g had been reached. Meier and Willis reduced the feeding frequency to 6 times daily at 6 weeks of age.

A variety of 'bottles' have been used to provide milk, including eye-drop dispensers, (Pet) bottles and nipples, and syringes (Cartmill et al. 1979; Brockman et al. 1987b). Brockman et al. (1987b) began to offer milk formula in dishes when babies had reached 2 months of age. In all cases, milk was provided *ad libitum* at each feed.

Accommodation

At San Diego Zoo, nest boxes made of opaque acrylic plexiglass with a 1×0.5 metre floor area and a height of 0.6 metres, were provided in each breeding cage. These nest boxes were divided completely by a removable clear plastic partition which allowed the keepers to observe animals in either compartment by opening the door to the unoccupied compartment. The partition also enabled the separation and locking of the mother into the opposite compartment from the babies using bait, thus permitting access to the litter for inspection and weighing. The nest box floors were heated by coiled hot water pipes (Brockman et al. 1987a). Prior to parturition a soft substrate (washable indoor-outdoor carpet) was provided in the nest box and changed 2 to 3 times a week (Brockman et al. 1987b).

Infants taken for hand-rearing should be kept in incubators at about 32 °C and 50–60 per cent humidity (Meier and Willis 1984), for the first month, and the temperature gradually reduced to room temperature (about 25 °C) prior to removing the babies to nursery cages. In these early stages it has been found best to house infants individually to prevent them sucking at each other's extremities, and also because they are easier to keep clean (Brockman et al. 1987b). As ruffed lemurs do not cling to their mothers but spend their early days in a nest, there is no rationale for providing a surrogate of the sort used in many other primate species. Brockman et al. (1987b) recommended the transfer of infants to cages with other young at 1.5 to 2 months of age.

Infant management notes

Warm baths (with a mild disinfectant added to the water) have been used to bring the temperature of hypothermic infants to 37 °C (Brockman et al. 1987b). It may be necessary to provide supplementary heat sources, such as hot-water bottles or heat pads, in the incubator for weak babies. Meier and Willis (1984) treated the umbilicus once daily with a dilute iodine solution for three days. Infants should be weighed before feeding each morning to monitor progress. Because mother-reared infants are deposited in the nest and not carried by the mother, there is a greater opportunity to monitor the health and growth of these babies than with other species. Cartmill et al. (1979) corrected mild espisodes of constipation by administering 1 or 2 drops of mineral oil, and 'burped' babies after meals to prevent colic.

Hand-reared babies are not easy to keep clean in incubators and Meier and Willis (1984) recommended regularly bathing the infants using a hypoallergenic soap and rinsing them thoroughly afterwards. Although the infants do not need a surrogate mother, Brockman et al. (1987b) considered that grooming the infants for 10 minutes each day may be beneficial to behavioural development.

Physical development

Neonatal ruffed lemurs are covered with fur and their eyes are open at birth (Foerg 1982). The sequence of dental eruption is thought to be upper and lower first molars, lower second incisors, upper third incisors, upper second incisors, lower canines, upper and lower second molars, upper canines, lower second premolars, upper second premolars, upper and lower fourth premolars, upper and lower third molars, and upper and lower third premolars (Swindler 1976). The testes descend from the inguinal canal to the scrotum at 16 to 22 months of age (Foerg 1982).

Behavioural development

At birth, infant *Varecia* are unable to cling to their mothers like other lemur species. However, development is rapid and they are capable of independent locomotion by 2 months of age (Boskoff 1977). During the first week the mother spends over 80 per cent of her time with the babies (Klopfer and Dugard 1976). At birth, the babies can crawl and seek the nipple (Foerg 1982). By day 8 they are able to support their bodies off the ground, walk, and appear to be able to focus on objects. From birth, babies make a short nasal moaning vocalization when hungry and, during the second week begin to purr like cats when being groomed or after feeding. At 16 days of age they can move with a fast hopping gait (all the above notes are from Cartmill *et al.* 1979).

Klopfer and Dugard (1976) observed two females that first carried their babies (using their mouths) out of the nest box at 21 and 29 days of age but soon returned them. The babies made their first independent excursions from the nest boxes at 25 and 32 days of age. Brockman *et al.* (1987b) recorded babies beginning to leave the nest box at 3 weeks of age and also noted that the fathers take little interest in the babies during the first 2 weeks but later played with and groomed them.

Perineal marking behaviour was first observed in 2-month-old infants and the adult choral call was first made at 3 months of age (Cartmill *et al.* 1979).

Disease and mortality

In the *International studbook*, Brockman (1986) reported that 118 out of 481 (25 per cent) of *V.v. variegata* and 34 out of 164 (21 per cent) of *Vv. rubra* were stillborn or died within 30 days of birth. The total deaths during the first year were, however, not much greater — 141 out of 507 (28 per cent) *V.v. variegata*, and 38 out of 164 (23 per cent) *V.v. rubra*. Thus, after the first month, mortality was low. Of 51 ruffed lemurs of both subspecies entering the nursery at San Diego between 1972 and 1981, 31 (61 per cent) died and the rest were successfully hand-reared. Low birth weight and prematurity, as indicated by the histological appearance of the renal cortices, were important factors in neonatal mortality (Benirschke and Miller 1981). Maternal neglect is also an important factor, and was considered by Brockman *et al.* (1987a) to be usually secondary to infant hypothermia, low birth weight, large litters, or congenital abnormalities (Benirschke *et al.* 1981).

Brockman *et al.* (1987b) reported that the survival of low birth weight infants (70 g or less) was much improved when they were taken for hand-rearing. The smallest infant successfully reared by Meier and Willis (1984) weighed only 60 g.

Preventative medicine

At San Diego, where relatively large numbers of ruffed lemurs are hand-reared, faecal samples were collected and screened for *Salmonella* once a week. When the infants were 6 months old both they and the mother were given a health check, including routine haematological examination, tuberculin test, and radiographic examination (Brockman *et al.* 1987b).

Indications for hand-rearing

Brockman *et al.* (1987b) suggested that if all seems to be going well the babies should be inspected and weighed about 6 to 8 hours after birth. Meier and Willis (1984) recommended removal of infants weighing less than 85 g for hand-rearing, with the exception of singletons weighing 80–85 g which may be best left with the mother and monitored carefully. If the litter size exceeds two it may be necessary to remove one or more of the babies for hand-rearing. Infants should also be taken for hand-rearing if their mothers show no interest in them, or show extremely agitated or aggressive behaviour. Neglected hypothermic babies which are otherwise healthy may be successfully reintroduced to their mothers after temporary removal for warming and administration of about 3 ml oral glucose/electrolyte solution (Meier and Willis 1984). Cross-fostering neglected infants to other females that have very recently lost their own litters should be considered as an alternative to hand-rearing (Shideler and Lindburg 1982).

Reintegration

It may be possible, as mentioned above, to reintroduce very young infants to their own mother or to a foster-mother, if they were abandoned because of hypothermia and regain their strength after being warmed. Hand-reared young should be kept in groups from 1.5 to 2 months of age and introduced to adults at 4 to 6 months of age (Brockman *et al.* 1987*b*). These authors first allowed visual, auditory, and olfactory contact between the weanlings and an adult pair through an introduction screen for periods of 1 to 2 weeks (or longer if there were still signs of aggression), after which the screen was removed. They noted low-level aggression after introductions but this declined after about a week, and no severe injuries occurred. The infants were left with the adults through the birth of the following, year's litter. Meier and Willis (1984) considered that the behaviour of hand-reared and mother-reared infants was similar, and recorded that hand-reared ruffed lemurs have reproduced successfully and reared their young.

References

Benirschke, K. and Miller, C.J. (1981). Weights and neonatal growth of ring-tailed lemurs (*Lemur catta*) and ruffed lemurs (*Lemur variegatus*). *Journal of Zoo Animal Medicine*, **12**, 107–11.

Benirschke, K., Kumamoto, A.T. and Bogart, M.H. (1981). Congenital anomalies in *Lemur variegatus*. *Journal of Medical Primatology*, **10**, 38–45.

Bogart, M.H., Cooper, R.W., and Benirschke, K. (1977*a*). Reproductive studies of black and ruffed lemurs *Lemur macaco macaco* and *L. variegatus spp*. *International Zoo Yearbook*, **17**, 177–82.

Bogart, M.H., Kumamoto, A.T., and Lesley, B.K. (1977*b*). A comparison of the reproductive cycle of three species of lemur. *Folia Primatologica*, **28**, 134–43.

Boskoff, K.J. (1977). Aspects of reproduction in ruffed lemurs (*Lemur variegata*). *Folia Primatologica*, **28**, 241–50.

Brockman, D.K. (1986). *International studbook of the ruffed lemur Varecia variegata* (2nd edn). Zoological Society of San Diego.

Brockman, D.K., Willis, M.S., and Karesh, W.B. (1987*a*). Management and husbandry of ruffed lemurs, *Varecia variegata*, at the San Diego Zoo. I. Captive population, San Diego housing and diet. *Zoo Biology*, **6**, 341–7.

Brockman, D.K., Willis, M.S., and Karesh, W.B. (1987*b*). Management and husbandry of ruffed lemurs, *Varecia variegata*, at the San Diego Zoo. II. Reproduction, pregnancy, parturition, litter size, infant care and reintroduction of hand-raised infants. *Zoo Biology*, **6**, 343–69.

Buss, D.H., Cooper, R.W., and Wallen, K. (1976). Composition of lemur milk. *Folia Primatologica*, **26**, 301–5.

Cartmill, M., Brown, K., Eaglen, R., and Anderson, D.E. (1979). Hand-rearing twin ruffed lemurs *Lemur variegatus* at the Duke University Primate Center. *International Zoo Yearbook*, **19**, 258–61.

Foerg, R. (1982). Reproductive behaviour in *Varecia variegata*. *Folia Primatologica*, **38**, 108–21.

Hick, U. (1976). The first year in the new Lemur house at Cologne Zoo. *International Zoo Yearbook*, **16**, 141–52.

IUCN (1990) *1990 IUCN Red list of threatened animals*, pp. 9. IUCN, Gland, Switzerland.

Jolly, A. (1986). Lemur survival. In *Primates. The road to self-sustaining populations* (ed. K. Benirschke), pp. 71–98. Springer-Verlag, New York.

Klopfer, P.H. and Dugard, J. (1976). Patterns of maternal care in lemurs. III. *Lemur variegatus*. *Zeitschrift für Tierpsychologie*, **40**, 210–20.

Lindsay, N.B.D. (1977). Notes on the taxonomic status and breeding of the ruffed lemur *Lemur (Varecia) variegatus*. *Dodo*, **14**, 65–9.

Meier, J.E. and Willis, M.S. (1984). Techniques for hand-raising neonatal ruffed lemurs (*Varecia variegata* and *Varecia variegata rubra*) and a comparison of hand-raised and maternally raised animals. *Journal of Zoo Animal Medicine*, **15**, 24–31.

Pollock, J.I. (1986). The management of prosimians in captivity for conservation and research. In *Primates. The road to self-sustaining populations* (ed. K. Benirschke), pp 269–88. Springer-Verlag, New York.

Shideler, S.E. and Lindburg, D.G. (1982). Selected aspects of *Lemur variegatus* reproductive biology. *Zoo Biology*, **1**, 127–34.

Swindler, D.R. (1976). *The dentition of living primates*, pp. 17–34. Academic Press, London.

Tattersall, I. (1982). *The primates of Madagascar*, pp. 67–73. Columbia University Press, New York.

Taylor, S.H. and Bietz, A.D. (Ed.) (1985). *Infant diet/care notebook*. American Association of Zoo Parks and Aquariums, Wheeling, West Virginia.

Lesser mouse lemur

2 Lesser mouse lemur

Species
The lesser mouse lemur   *Microcebus murinus*

ISIS No.   1406001

Status, subspecies and distribution
The lesser mouse lemur is found throughout Madagascar except for the central plateau, and is relatively abundant and not threatened at present (Wolfheim 1983; Jolly 1986). It is found in a variety of forest and woodland habitats that contain relatively dense foliage (Wolfheim 1983; Martin 1972).

The species is kept in zoos and laboratories and Pollock (1986) estimated the captive population to be 147 in 1985, based on ISIS and *International Zoo Yearbook* data. There were 170 ISIS-registered individuals as of December 1988 (ISIS 1989). They were first bred in captivity in 1957 (Jones 1986).

Sex ratio
In captivity, the ratio of males to females at birth was found by Glatston (1979) to be 1.2 to 1. Perret (1982) consistently found that more males were born than females (of 76 babies, 49 were male and 27 female: a ratio of 1.8:1), but observed that sex ratio appeared to be related to management conditions. The ratio of males to females was higher in babies born to females kept in groups than those kept with a male only.

Social structure
Lesser mouse lemurs form population nuclei with a core group with a high ratio of females to males, and a surrounding fringe of males which probably have no access to females for breeding. The females nest in groups in hollow trees or leaf-nests and the males live singly or in pairs. These groups remain fairly stable (Martin 1972, 1975). Social group structure influences reproduction in captivity. When several females are kept together, repeat oestrous cycles are depressed, there is an increase in rates of fetal resorption and abortion, more triplet litters occur, the ratio of male to female births increases, and neonatal mortality increases, compared with the performance of pairs (Perret 1982).

Breeding age
Animals become sexually mature in the breeding season after their birth. Females can give birth at 13 months of age, but both females and males usually start breeding later than this due to social interactions in the group (Glatston 1979).

Longevity
Manley (1966) recorded a lifespan of 11 years in one individual.

Seasonality
There is a clear seasonal distribution of births during the period of increasing day-length. Births occur between March and July in Europe and between September and March in the wild in Madagascar (Bourlière *et al.* 1961; Martin 1971, 1972). Two litters can be produced in one breeding season (Martin 1971; Pollock 1986).

Females are seasonally polyoestrus and the first cycle occurs about two months after the shortest day. There are three cycles per season if the female fails to conceive, but a decline in the incidence of repeat cycles has been observed when pairs are kept together (Perret 1982). Perret (1982) found that the first, second, and third oestrous periods were centred around April, June, and July respectively.

The vulva is imperforate except at oestrus and parturition (Tattersall 1982; Glatston 1981). During pro-oestrus the vulva swells and becomes pink. After 10 days the vaginal opening is visible and the vulva turns white. Copulation usually occurs 2 or 3 days after the vagina opens, and a week later the vagina closes (Bourlière *et al.* 1961; Martin 1972; Glatston 1979). The cycle length is 40 to 60 days (Glatston 1979).

Males show a seasonal cycle in testis size. The testes reach maximum size at the time the females begin oestrors cycles and decline after 4 to 6 months when spermatogenesis ceases (Glatston 1979).

Gestation

The gestation period is 54 to 69 days with a mean of 61 to 63 days (Bourlière *et al.* 1961; Martin 1972; Glatston 1979; Van Horn and Eaton 1979). It is thought to be 57 to 59 days in the red subspecies *M.m. rufus* (Glatston 1979).

Pregnancy diagnosis

Pregnancy can be diagnosed by abdominal palpation from about one month post-conception (Martin 1972). Ultrasonography or radiography could also be used. Nesting behaviour is observed shortly before birth (Glatston 1979).

Birth

In this nocturnal primate, births occur in the early daylight hours. When in labour the female sits with her legs apart and the parturition process takes 45 to 60 minutes, with as little as 5 minutes between infants. The placentae are eaten (Glatston 1979).

Litter size

The range is from 1 to 3, but twins are usual (Martin 1972; Glatston 1979; Van Horn and Eaton 1979).

Adult weight

Both sexes average about 60 g when adult, with a range of 50 to 100 g (Martin 1972; Hladik *et al.* 1980). There is a seasonal cycle in body weight with a minimum during the breeding season and a peak during the winter, associated with the storage of subcutaneous fat over the body and in the tail (Petter-Rousseaux 1974, 1980; Glatston 1979; Hladik *et al.* 1980). In captivity, animals may become obese and attain weights up to 150 g (Glatson 1979).

Neonate weight

Healthy neonates weigh between 5.5 and 7.0 g (Glatston 1979). Birth weight is inversely related to litter size. Perret (1982) found the mean weight of babies of single, twin, and triplet litters to be 8.45, 7.07, and 6.31 g respectively.

Adult diet

Wild mouse lemurs eat fruit and insects (Martin 1972; Hladik *et al.* 1980). In captivity they have been successfully kept on diets of mixed fresh and dried fruit, nuts (almonds and hazelnuts), bread, cooked rice, invertebrates such as locusts, crickets, and mealworms, and baby rats and mice. Vitamin and mineral supplements, such as Abidec and Vionate, have been added to the food (Bourlière *et al.* 1961; Martin 1972, 1975; Glatston 1979; Hladik 1979; Petter-Rousseaux and Hladik 1980; Poliak 1981). Pollock (1986) cautioned against feeding excess vitamin D_3 because these animals are rarely exposed to sunlight and their diet in the wild probably contains rather little. It is possible that their vitamin D_3 requirements are low and they may be sensitive to high dietary levels of this vitamin. Pollock (1986) suggested that lights should include wavelengths less than 310 nm, and that dietary vitamin D levels should not exceed 2000 IU/kg.

Pollock (1986) estimated the protein consumption of the lesser mouse lemur to be 0.9 to 2.6 g/d with a mean of 1.5 g/d.

Adult energy requirements

In the wild, mouse lemurs become torpid during the winter. In captivity, there is seasonal variation in food intake from 34g/d (of fresh food) during winter, when body weight is highest, to 72, 82, and 74 g/d in spring, summer, and autumn respectively (Petter-Rousseaux and Hladik 1980). These values correspond to a variation in energy intake from 7 kcal/d per 100 g body weight in winter to 35 kcal/d per 100 g body weight in summer (Hladik *et al.* 1980; Petter-Rousseaux 1980; Poliak 1981), or roughly 40 and 170 kcal/d per $kg^{0.75}$ respectively. These measurements are consistent with the finding that these animals have a relatively low metabolic rate for mammals of their size (Pollock 1986).

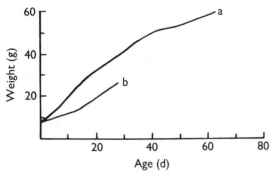

Fig. 2.1. Typical growth curves of (a) mother-reared and (b) hand-reared lesser mouse lemurs. From Martin (1975), Glatston (1981), and Perret (1982).

Growth

There is considerable variation in growth rates amongst young lesser mouse lemurs depending on the environment during growth. Babies hand-reared by Glatston (1981) grew more slowly than those reared by their mothers, reaching a mean weight of about 28 g at 30 days of age compared with 40 g in the mother-reared group. However, the mean weight of mother-reared babies recorded by Martin (1975) was about 30 g at 30 days of age. Perret (1982) observed that the growth rate of babies was influenced by the density of the maternal group; growth rate was slower in babies born to mothers kept in larger groups.

Under good conditions, adult summer body weight is reached when 60 days old and a typical growth curve based on data from Martin (1975), Glatston (1981), and Perret (1982) is shown in Fig. 2.1. The mean growth curve of hand-reared individuals based on data from Glatston (1981) is also shown in Fig. 2.1.

During the first 30 days, growth rate averages 0.6 to 1.0 g/d.

Milk and milk intake

As far as we are aware there are no data on the milk composition of this species. Martin (1972) and Glatston (1981) used human milk replacers, but as the growth rate of this species is relatively rapid compared with that of man, after body weight differences are taken into account (Kirkwood 1985), a milk replacer with a higher protein concentration (for example, Primilac) might be more appropriate.

There are no detailed accounts of the milk intake during growth but Glatston (1981) reported that the quantity of milk taken per feed increased from 0.5 ml during the first few days to 1.5 to 2.0 ml by 3 weeks. As she was providing 8 feeds a day, this corresponds to total intakes of 4 ml/d rising to about 14 ml/d at 3 weeks, and (assuming a milk energy density of 0.67 kcal/ml) this represents energy intakes of 2.7 kcal/d rising to 9.4 kcal/d. Since the mean weight of Glatston's hand-reared individuals was about 8 g at 1 day old and about 20 g at 3 weeks, these energy intakes correspond to about 100 kcal/d per metabolic weight$^{0.75}$ initially, rising to about 260 kcal/d per metabolic weight at 3 weeks of age.

Lactation and weaning

Babies have been observed to suck within 12 minutes of birth (Glatston 1979), and sucking bouts last 20 minutes or longer. Babies start eating solid food at about three weeks of age, and lactation ceases at about 45 days of age (Bourlière et al. 1961; Martin 1972; Glatston 1981).

Feeding

Glatston (1981) used a small pipette or eye-dropper without a teat for feeding. She fed the babies at roughly 3 hour intervals, including a night feed, but Martin (1972) did not feed between late night and early morning. After 3 weeks the number of feeds was reduced to 5, and to 2 per day at 4 weeks of age. Solid foods (including mashed banana and chopped mealworms) have been offered and taken from about 3 weeks of age (Martin 1972; Glatston 1982).

Accommodation

Glatston (1982) recommended that newborn babies should be kept at 30 °C, and observed that they failed to gain weight if kept at lower temperatures. She kept them in lined glass beakers suspended in a water bath but, where the resources are available, an incubator would be simpler.

Initially, babies should be kept separately as they tend to suck each other's genitals (Glatston 1982).

Infant management notes

Stimulation of the perineum with a damp cloth may be necessary to promote urination and defaecation in young babies. Both occur at each feed (i.e., at least 8 times a day). Defaecation becomes spontaneous when the infants begin taking solid food (Martin 1972; Glatston 1982).

Physical development

The eyes open 2 to 4 days after birth. At birth, the dorsum is covered with soft fur but the abdomen is naked and the skin transparent enough to be able to see if the stomach contains milk or not. The testes are inguinal at birth and descend into the scrotum at 4 to 5 days of age.

The sequence of dental eruption in the upper jaw is first the first molars, and second, and third incisors, next the second molars, then the canines, then the second premolars, followed by the third molars, the fourth premolars and the third premolars. In the lower jaw the sequence is: first molars, second molars, second incisors, third incisors, canines, second premolars, third molars, fourth premolars, and third premolars (Schwartz 1975; Tattersall 1982).

Behavioural development

From birth, babies are very active and are able to cling to their mother's fur and support their own weight when suspended. They are unable to raise their bodies off the ground until the second week. At about 7 days of age, they can begin to crawl and climb, but they do not attempt to jump until about 4 weeks old. At 10 to 13 days old, the babies begin to show play behaviour and adopt the adult sleeping position, curled up with their tails over their snouts. Mother-reared babies spend the first 3 weeks within the nest, but from 4 weeks old they spend most of their time outside the nest, and by 3 months they are more or less independent (Glatston 1979).

Disease and mortality

Perret (1982) isolated all pregnant females for parturition and lactation but found that infant mortality (to 5 months of age) was related to the previous system of management, being higher amongst infants born to females that had been kept in groups than in those born to females that had been kept in pairs. This may be related to the fact that group-kept females tended to have larger litters with lower individual birth weights. The parity of the dam did not seem to affect infant mortality rate.

Glatston (1982) reported the successful hand-rearing of five out of eight infants, and the three that died all weighed less than 4.5 g at birth. She reported few problems in hand-reared infants but diarrhoea occurred on occasions and was treated by diluting the milk formula. Still births and cannibalism have been observed and have been considered to be associated with stress resulting from handling, disturbance, and overcrowding (Martin 1975; Glatston 1979). Pneumonia and enteritis have been identified as causes of death among infants (Glatston 1979).

Preventative medicine

From the above it is clear that the social environment of the dam has an important influence on infant mortality and it is better to keep the animals in breeding pairs than in multi-female groups. As far as we are aware there is no special information on disease prevention in the young of this species.

Indications for hand-rearing

Hand-rearing should be considered for abandoned or orphaned babies, and the chances of survival of those (especially triplets) of particularly low birth weight may be improved if taken for hand-rearing.

Reintegration

Glatston (1982) reintroduced hand-reared babies to the colony when 50 to 55 days old. She noted that 'none of the detrimental effects which hand-rearing induces in the reproductive or parental behaviour of various other primate species, is apparent in the mouse lemur'. Some of her hand-reared animals subsequently became the most prolific breeders in the colony and she considered that their friendly and tractable nature had advantages. The behaviour of hand-reared babies did not appear to differ significantly from that of those reared naturally (Glatston 1979). Martin (1975) had previously observed that imprinting on man and subsequent problems with breeding had not been encountered in this species.

References

Bourlière, F., Petter-Rousseaux, A., and Petter, J. (1961). Regular breeding in captivity of the lesser mouse lemur (*Microcebus murinus*). *International Zoo Yearbook*, 3, 24–5.

Glatston, A.R. (1979). Reproduction and behaviour of the lesser mouse lemur (*Microcebus murinus*, Miller 1977) in captivity. D.Phil. thesis. University of London.

Glatston, A.R. (1981). The husbandry, breeding and hand-rearing of the lesser mouse lemur *Microcebus murinus* at Rotterdam Zoo. *International Zoo Yearbook*, 21, 131–7.

Hladik, C.M. (1979). Diet and ecology of prosimians. In *The study of prosimian behaviour* (ed. G.A. Doyle and R.D. Martin), pp. 307–57. Academic Press, New York.

Hladik, C.M., Charles-Dominique, P., and Petter, J.J. (1980). Feeding strategies of five nocturnal prosimians in the dry forest of the west coast of Madagascar. In *Nocturnal Malagasy primates: ecology, physiology and behaviour* (ed. D.M. Rumbaugh), pp. 41–73. Academic Press, New York.

ISIS (1989). *Species distribution report abstract: mammals*. ISIS, Minnesota.

Jolly, A. (1986) Lemur survival. In *Primates. The road to self-sustaining populations* (ed. K. Benirschke), pp. 71–98. Springer-Verlag, New York.

Jones, M.L. (1986). Successes and failures of captive breeding. In *Primates. The road to self-sustaining populations* (ed. K. Benirschke), pp. 251–60. Springer-Verlag, New York.

Kirkwood, J.K. (1985). Patterns of growth in primates. *Journal of Zoology*, 205, 123–36.

Manley, G.H. (1966). Prosimians as laboratory animals. *Symposia of the Zoological Society of London*, 17, 11–39.

Martin, R.D. (1971). A review of the behaviour and ecology of the lesser mouse lemur (*Microcebus murinus*, Miller 1977). In *Comparative ecology and behaviour of primates* (ed. R.P. Michael and J.H. Crook), pp. 2–68. Academic Press, London.

Martin, R.D. (1972). A laboratory breeding colony of the lesser mouse lemur. In *Breeding primates* (ed. W.I.B. Beveridge), pp. 161–71. S. Karger, Basel.

Martin, R.D. (1975). Breeding tree shrews (*Tupaia belangeri*) and mouse lemurs (*Microcebus murinus*) in captivity. *International Zoo Yearbook*, 151, 35–41.

Perret, M. (1982). Influence du groupement social sur la reproduction de la femelle de *Microcebus murinus* (Miller, 1977). *Zeitschrift für Tierpsychologie*, 60, 47–65.

Petter-Rousseaux, A. (1974). Photoperiod, sexual activity and bodyweight variation of *Microcebus murinus*. In *Prosimian biology* (ed. R.D. Martin, G.A. Doyle and A.C. Walker), pp. 169–79. Duckworth, London.

Petter-Rousseaux, A. (1980). Seasonal activity rhythms, reproduction and body weight variations in five sympatric nocturnal prosimians in simulated light and climatic conditions. In *Nocturnal Malagasy primates: ecology, physiology and behaviour*, pp. 137–52. Academic Press, London.

Petter-Rousseaux, A. and Hladik, C.M. (1980). A comparative study of food intake in five nocturnal primates in simulated climatic conditions. In *Nocturnal Malagasy primates: ecology, physiology and behaviour*. (ed. D.M. Rumbaugh). Academic Press, London.

Poliak, S. (1981). L'alimentation des lemuriens en captivité. Thèse pour le Doctorat Vétérinaire. Ecole Nationale Vétérinaire de Toulouse.

Pollock, J.I. (1986). The management of prosimians in captivity for conservation and research. In *Primates. The road to self-sustaining populations* (ed. K. Benirschke), pp. 269–88. Springer-Verlag, New York.

Schwartz, J.H. (1975). Development and eruption of the premolar region of prosimians and its bearing on their evolution. In *Lemur biology* (ed. I. Tattersall and R.W. Sussman), pp. 41–64. Plenum Press, New York.

Tattersall, I. (1982). *The primates of Madagascar*. Columbia University Press, New York.

Van Horn, R.N. and Eaton, G.G. (1979). Reproductive physiology and behaviour in prosimians. In *The study of prosimian behaviour* (ed. G.A. Doyle and R.D. Martin), pp. 79–122. Academic Press, New York.

Wolfheim, J.H. (1983). *Primates of the world*, pp. 71–4. University of Washington Press, Seattle.

Senegal or lesser bush-baby

3 Senegal or lesser bush-baby

Species

The Senegal or lesser bush-baby
Galago senegalensis

ISIS No.   1406004006004

Status, subspecies, and distribution

There are nine subspecies, distributed throughout much of Subsaharan Africa, excluding the Zaïre river basin and the southern tip of the continent (Doyle 1974; Wolfheim 1983). Lesser bush-babies occur mainly in savanna and woodland habitats, and are common in some areas (Wolfheim 1983). The species was classified as not threatened (Lee *et al.* 1988).

The captive population in 1985 was estimated by Pollock (1986) to be 353, and there were 166 ISIS-registered individuals as of December 1988 (ISIS 1989). The first recorded breeding in captivity was at the Zoological Society of London in 1855 (Jones 1986). The captive population appears to be self-sustaining (Flesness 1986).

Sex ratio

Of 60 infants born at the Duke University Primate Center, 67 per cent were male (Izard and Simmons 1986*b*).

Social structure

Galagos are found as singles and in groups of up to nine individuals (Manley 1966), but the usual structure consists of a pair with two to four offspring (Bearder and Doyle 1974).

Breeding age

Conceptions have been recorded in animals of just 120 days old, but age at first conception is more usual at about 200 days (Doyle 1979). Females housed with peer males reach puberty at a later stage (about 1 year old) than when they are kept with adult males, in which case they become sexually mature at 200 to 300 days old (Izard and Simons 1986*b*).

Longevity

Lesser galagos have lived up to 10 years in captivity (Manley 1966), but the maximum lifespan is probably considerably longer.

Seasonality

In the wild, the lesser bush-baby is thought to have a seasonal breeding pattern with two birth seasons each year. Southern subspecies appeared to have birth seasons from October to November and from January to March (Bearder and Doyle 1974), and northern subspecies have birth seasons from February to March and May to June (Van Horn and Eaton 1979). However, there is some variation within the ranges of the subspecies (Butler 1967). In captivity, there are typically two births a year (Manley 1966; Doyle *et al.* 1967; H. Gucwinska & A. Gucwinska 1968), but the seasonal pattern is lost and births can occur in any month (Darney and Franklin 1982).

The duration of the oestrous cycle was found to vary between 29 and 39 days, and oestrus lasted from 4 to 7 days (Darney and Franklin 1982).

Gestation

The duration of gestation was found in a sample of 61 pregnancies to be 124 days (Izard and Simons 1986*a*).

Pregnancy diagnosis

Pregnancy can be detected 30 to 40 days after conception by abdominal palpation (Izard and Simons, 1986*b*), and pregnant females show a weight gain in the last month before parturition (Doyle *et al.* 1967).

Birth

Births mostly occur during the day (H. Gucwinska and A. Gucwinska 1968; Darney and Franklin 1982), but have been observed to occur at night among the *G.s. moholi* subspecies (Doyle *et al.* 1967). Nesting behaviour is sometimes but not always shown in the 36 hours preceding parturition (Doyle *et al.* 1967).

Litter size

The litter size is usually one or two, and twins are more common among multiparous females. Triplets are exceptional (Doyle *et al.* 1971; Bearder and Doyle 1974; Izard and Simons 1986*a, b*).

Adult weight

Doyle (1979) reported mean weights of adult males and females at 235 g and 195 g respectively, but weights vary considerably within the range of 150 to 300 g between individuals even within the same colony (Bearder and Doyle 1974; Doyle 1979; Izard and Simons 1986*a, b*).

Neonate weight

The mean weight of 28 male babies was found to be 11.9 g, with a range from 8.6 to 15.5 g, and that of 28 females was 11.7 g with a range from 8.5 to 14.8 g (Doyle 1979).

Adult diet

Bush-babies are insectivorous but also eat *Acacia* gum, fruit, and nectar (Doyle 1974). In captivity, they have been maintained on diets of insects (mealworms and crickets) and chopped fruit (Doyle 1974), and they will also take pelleted primate diet and cat food (Darney and Franklin 1982; Izard and Simons 1986*b*), and inclusion of a nutrionally-balanced pellet is recommended.

Adult energy requirements

There appears to be no specific information on the daily requirement.

Growth

Information on growth has been provided by Brown (1979), Doyle (1979), and H. Gucwinska and A. Gucwinska (1968). The growth rate during the first month is approximately linear and equal to about 1.5 g/d (Fig. 3.1). Thereafter, growth rate declines gradually and adult weight is approached at about one year of age, although there is evidence of continued weight gain during the second year (Fig. 3.2).

Milk and milk intake

As far as we are aware the milk of the Senegal bush-baby has not been analysed. The milk of the thick-tailed galago, *Galago crassicaudatus*, was reported by Pilson and Cooper (1967) to contain 2.74 to 4.60 per cent fat, 4.18 to 6.59 per cent protein, and 4.86 to 5.14 per cent lactose. The energy content was therefore probably close to 0.76 kcal/ml and the composition of the dry matter was probably close to 25 per cent fat, 36 per cent protein, and 33 per cent lactose. The milk of *G. crassicaudatus*, therefore, has a considerably higher protein content than that of most New World primates, cercopithecines, and apes. It is likely that the milk of *G. senegalensis* is similar in composition. In spite of this, babies have been reared on the human milk replacer Goldcap SMA (Brown 1979), but milk replacers with a higher protein concentration, such as Lactol or Lamlac, might prove more suitable 'off-the-shelf' milk substitutes. Brown (1979) added glucose to the SMA but found that this caused diarrhoea, perhaps because the natural milk of this species contains considerably less carbohydrate than SMA even before supplementation.

Lactation and weaning

Infants suck whilst lying on their backs beneath their mother and clinging to her head with both hind feet (H. Gucwinska and A. Gucwinska 1968). The first solid foods are taken at 17 to 20 days, and are eaten regularly at 1 month of age. Weaning is largely completed by 70 to 80 days but sucking has been observed up to 100 days (Manley 1966; H. Gucwinska and A Gucwinska 1968; Doyle 1979).

Feeding

Brown (1979) hand-reared twin bush-babies using a 1 ml pipette fitted with a bicycle valve rubber as

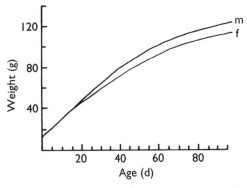

Fig. 3.1. Average growth curves of male (m) and female (f) Senegal bush-babies to 90 days of age. From Doyle (1979).

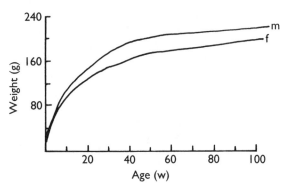

Fig. 3.2. Average growth curves of male (m) and female (f) Senegal bush-babies to 2 years of age. From Doyle (1979).

a teat. These babies were fed 9 times a day for the first 3 weeks, then the frequency was gradually reduced to 4 feeds per day by 6 to 7 weeks of age. Milk to which Farex had been added was made available in the cage from an early stage and the babies increasingly fed on this as the frequency of hand-feeds was reduced.

Insects, such as moths, craneflies, lacewings, and mealworms, were taken from about 3 weeks of age, and fruit was taken from 6 weeks of age.

Accommodation

No special accommodation has been devised for infant galagos.

Infant management notes

Brown (1979) found that no stimulation was necessary for urination and defaecation.

Physical development

The newborn *G. senegalensis* is covered with grey fur and the eyes are open. The sequence of premolar eruption in both jaws is second, fourth, third for both deciduous and permanent sets (Schwartz 1975).

Behavioural development

Bush-babies are quite mobile from birth and can crawl and pull themselves into the sucking position. They can see and hear, and take an interest in the environment. They can walk with their bellies clear of the ground, and can climb on to their mother's back. However, when changing nests or retrieving her babies the mother carries them with her mouth (H. Gucwinska and A. Gucwinska 1968). The babies remain in the nest until about 10 days of age when they begin to explore outside. They are able to run fast by 2 weeks of age, can jump, and begin climbing branches at 3 weeks of age. By 4 to 6 weeks infants spend most of the night away from their mothers (Doyle 1979).

Disease and mortality

Quite high rates of neonatal mortality have been reported, for example, from 25 per cent (Doyle 1974) to 73 per cent (Izard and Simons 1986a). Izard and Simons (1986b) reduced substantially the rate of infant mortality (from 65 to 16 per cent) in a captive colony by isolating pregnant females 5 days prior to parturition. This reflects the situation in the wild, where females with babies distance themselves from the group until the babies are about 2 weeks old.

Preventative medicine

No specific data are available as far as we are aware.

Indications for hand-rearing

There are no specific indications. One weak baby was left with its mother but was given supplemental feeding and survived (H. Gucwinska and A. Gucwinska 1968).

Reintegration

No specific data are available. Hand-reared prosimians seem less prone to permanent behavioural abnormalities leading to inability to pair, mate, or rear young, than 'higher' primates.

References

Bearder, S.K. and Doyle, G.A. (1974). Ecology of bushbabies *Galago senegalensis* and *Galago crassicaudatus*, with some notes on their behaviour in the field. In *Prosimian biology* (ed. R.D. Martin, G.A. Doyle, and A.C. Walker), pp. 109–30. Duckworth, London.

Brown, C. (1979). Hand-rearing Senegal bushbabies *Galago senegalensis* at the Wildlife Breeding Centre. *International Zoo Yearbook*, **19**, 261–2.

Butler, H. (1967). Seasonal breeding of the Senegal galago (*Galago senegalensis senegalensis*) in the Nuba mountains of the Sudan. *Folia Primatologica*, **5**, 165–75.

Darney, K.J. Jr. and Franklin, L.E. (1982). Analysis of the oestrus cycle of the laboratory-housed Senegal galago (*Galago senegalensis*): natural and induced cycles. *Folia Primatologica*, **37**, 106–26.

Doyle, G.A. (1974). The behaviour of the lesser bushbaby. In *Prosimian biology* (ed. R.D. Martin, G.A. Doyle, and A.C. Walker), pp. 213–31. Duckworth, London.

Doyle, G.A. (1979). Development of behaviour in prosimians with special reference to the lesser bushbaby, *Galago senegalensis moholi*. In *The study of prosimian behaviour* (ed. G.A. Doyle and R.D. Martin), pp. 157–206. Academic Press, New York.

Doyle, G.A., Pelletire, A., and Bekker, T. (1967). Courtship, mating and parturition in the lesser bushbaby (*Galago senegalensis moholi*) under semi-natural conditions. *Folia Primatologica*, **7**, 169–97.

Doyle, G.A., Anderson, A., and Bearder, S.K. (1971). Reproduction in the lesser bushbaby (*Galago senegalensis moholi*) under semi-natural conditions. *Folia Primatologica*, **14**, 15–22.

Flesness, N.R. (1986). Captive status and genetic considerations. In *Primates. The road to self-sustaining populations* (ed. K. Benirschke), pp. 845–56. Springer-Verlag, New York.

Gucwinska, H. and Gucwinska, A. (1968). Breeding the Zanzibar galago. *International Zoo Yearbook*, **8**, 111–14.

ISIS (1989). *Species distribution report abstract: mammals*. ISIS, Minneapolis.

Izard, M.K. and Simons, E.L. (1986a). Infant survival and litter size in primigravid and multigravid *Galagos*. *Journal of Medical Primatology*, **15**, 27–35.

Izard, M.K. and Simons, E.L. (1986b). Management of reproduction in a breeding colony of bushbabies. In *Primate ecology and conservation*, Vol. 2 (J.G. Else and P.C. Lee), pp. 315–23. Cambridge University Press.

Jones, M.L. (1986). Successes and failures of captive breeding. In *Primates. The road to self-sustaining populations* (ed. K. Benirschke), pp. 251–60. Springer-Verlag, New York.

Lee, P.C., Thornback, J., and Bennett, E.L. (1988). *Threatened primates of Africa*. IUCN, Gland, Switzerland.

Manley, G.H. (1966). Prosimians as laboratory animals. *Symposia of the Zoological Society of London*, **17**, 11–39.

Pilson, M.E.Q. and Cooper, R.W. (1967). Composition of milk from *Galago crassicaudatus*. *Folia Primatologica*, **5**, 88–91.

Pollock, J.I. (1986). The management of prosimians in captivity for conservation and research. In *Primates. The road to self-sustaining populations* (ed. K. Benirschke), pp. 269–88. Springer-Verlag, New York.

Schwartz, J.H. (1975). Development and eruption of the premolar regions of prosimians and its bearing on their evolution. In *Lemur biology* (ed. I. Tattersall and R.W. Sussman), pp 41–64. Plenum Press, New York.

Van Horn, R.N. and Eaton, G.G. (1979). Reproductive physiology and behaviour of prosimians. In *The study of prosimian behaviour* (ed. G.A. Doyle and R.D. Martin), pp. 79–122. Academic Press, New York.

Wolfheim, J.H. (1983). *Primates of the world*, pp. 26–33. University of Washington Press, Seattle.

Potto

4 Bosman's potto

Species
Bosman's potto   *Perodicticus potto*

ISIS No.   1406004004001

Status, subspecies, and distribution
Five subspecies are recognized (Wolfheim 1983). It is a species which inhabits a wide range of forest habitats in western and central Africa from Guinea to western Kenya, in a belt extending 5 to 10 degrees on either side of the Equator (Wolfheim 1983). It was classified as not threatened by Lee *et al.* (1988).

Only 'a few' individuals are thought to be maintained in captivity (Pollock 1986).

Sex ratio
Sex ratios at birth seem to be close to 1:1. However, in the wild, adult females appear to outnumber males (Charles-Dominique 1977).

Social structure
With the exception of females with young, pottos are solitary. They sleep alone during the day, and forage by themselves during the night (Charles-Dominique 1977; Charles-Dominique and Bearder 1979).

Breeding age
Puberty is reached at about 12 to 18 months of age (Charles-Dominique 1977).

Longevity
Pottos are thought to live 15 years or more in captivity (Charles-Dominique 1984).

Seasonality
In the wild, pottos breed once a year, with births occurring between August and January. Males show an increase in testicular weight during the breeding season (Charles-Dominique 1974, 1977). However, this seasonal pattern may be lost in captivity if the environmental conditions are favourable all year round (Ioannou 1966).

Cycling in the female is characterized by the turgidity and opening of the vulva (Manley 1966). Interoestrus periods can vary from 34 to 47 days although a period of 37 to 39 days appears more common (Ioannou 1966; Manley 1966). Oestrus lasts 2.2 days on average (Ioannou 1966) and a post-partum oestrus is observed (Cowgill 1969).

Gestation
Gestation length is thought to be 193 days (Charles-Dominique 1977).

Pregnancy diagnosis
No specific data are available.

Birth
Females give birth on a branch. The mother eats the placenta and cleans the infant which clings to her body (Charles-Dominique 1977). Birth takes place at night (Crandall 1964).

Litter size
Pottos usually produce one offspring (Charles-Dominique 1974) although twins have been recorded (Cowgill 1974).

Adult weight
The average weight of adults is 1.1 kg (Charles-Dominique 1979) but weights of up to 1.58 kg have been recorded (Jewell and Oates 1969).

Neonate weight
Newborn pottos weigh approximately 50 g (Charles-Dominique 1977, 1984) although birth weights between 30 to 42 g have been recorded (Cowgill 1969). Twins, which failed to survive, weighed 37.7 g and 40.0 g (Cowgill 1974).

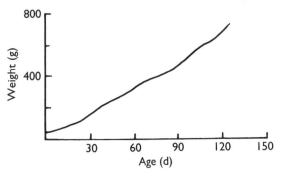

Fig. 4.1. Growth rate of one mother-reared male potto. From Jewell and Oates (1969).

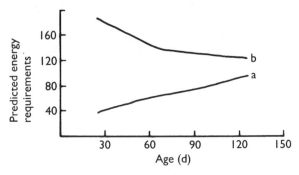

Fig. 4.2. Estimated daily energy requirements of infant pottos (a) total kilocalories per day (b) kilocalories per day per unit of metabolic weight (see text).

Adult diet

In the wild, pottos feed on fruits (65 per cent), animal prey (10 per cent), and gums (21 per cent), with some leaves and fungi. The animal prey consists of insects, such as ants (65 per cent), beetles (10 per cent), snails (10 per cent), caterpillars (10 per cent), and others. The stomach contents of adults were found to comprise 21 g fruit, 7 g gum, and 3.4 g of animal matter on average. Very occasionally, vertebrates, such as birds, bats, or rodents, are taken (Charles-Dominique 1974, 1977, 1979; Charles-Dominique and Bearder 1979). In captivity, pottos will take fruits and insects and are fond of bananas (Cowgill 1974).

Adult energy requirements

The basal metabolic rate of adult pottos has been measured at between 36 and 44 kcal/d per $kg^{0.75}$ (Pollock 1986), which is about half the average for mammals. The daily metabolizable energy intake of an adult is therefore likely to be about 60 to 80 kcal/d per $kg^{0.75}$, or about 75 kcal/d.

Growth

The growth of one mother-reared male potto from 1 to 126 days is shown in Fig. 4.1 (Jewell and Oates 1969). Another mother-reared baby showed a roughly linear weight gain from about 68 g when 4 days old to 675 g at 109 days old (Grand *et al.* 1984). Adult weight is reached between 8 and 14 months (Charles-Dominique 1977). The growth rate therefore appears to be about 5.5 g/d until half grown.

Milk and milk intake

The composition of potto milk has not been measured. Diluted cow's milk has been used (see below). In view of the rapid growth rate of the potto in relation to its adult size compared with cercopithecine and hominoid primates, the protein content of the milk is likely to be higher than that in human milk replacers.

No specific data on milk intake are available. In view of the low metabolic rate of this animal, the maintenance component of the energy requirement during growth is likely to be low in relation to body weight.

It is likely that the energy density of each gram of weight gained is about 2 kcal during the early stages of growth, and that about 3 kcal metabolizable energy are required for each gram of weight gain. As weight gain is close to 5.5 g/d, the energy required during growth in excess of that required for maintenance is probably about 18 kcal/d. If we assume that the maintenance requirement is 25 per cent higher during growth than when adult, i.e., 100 kcal/d per $kg^{0.75}$, it is possible to estimate the total daily energy requirement in relation to age. This is shown in Fig. 4.2.

Lactation and weaning

From days 3 to 8 onwards, babies are 'parked' at night while the mothers leave to forage alone. Thus the babies are only able to suck during the day. In the wild, infants are weaned at 120 to 180 days (Charles-Dominique 1977).

Feeding

Pottos have been hand-reared successfully (Crandall 1964; Walker 1968). The formula used in both reports was reconstituted evaporated milk (1 part milk to 2 parts water), supplemented with multivitamins (for example, Abidec drops at one drop per 15 ml of formula, Walker 1968). A doll's nursing bottle was used for feeding.

Walker (1968) fed one baby potto every 3 hours during the day and initially once during the night. The night feed was discontinued after about three weeks. The infant was gradually weaned on to banana, shredded beef, and diluted cow's milk. It took its first solid meal of mashed fruit at 5 weeks and its first live food (grasshoppers) at 6 months.

Accommodation

One infant was kept in a heated basket for the first month. The cotton-wool supplied as bedding was used as a surrogate mother and groomed by the infant.

Infant management notes

Stimulation of the perineal area was necessary to elicit urination and defaecation (Walker 1968; Cowgill 1974).

Physical development

The eyes are open at birth, but appear blue and misty (Walker 1968; Charles-Dominique 1977). The infant is covered with a sparse coat of white juvenile underhairs and longer guard hairs giving it a cream-coloured appearance with a light-brown dorsal stripe. The full adult pelage is developed over the first 6 months of life (Grand *et al.* 1964; Walker 1968). The cervical spines become prominent at 30 days of age (Grand *et al.* 1964) although Walker (1968) reported their presence at birth. The testes of the baby studied by Jewell and Oates (1969) had not descended at 4.5 months.

The deciduous and permanent premolars erupt in the following order in both jaws: second deciduous premolar, fourth deciduous premolar, third deciduous premolar, and second permanent premolar, fourth permanent premolar, third permanent premolar (Schwartz 1975), and the third molar erupts at 180 days (Charles-Dominique 1977). The adult weight is reached between 8 and 14 months (Charles-Dominique 1977).

Behavioural development

Newborn pottos cling to their mothers immediately after birth and seek the nipples. They are carried on their mothers' bodies until 3 to 8 days old. After that, they are 'parked' on a branch during the night whilst the mothers forage, and are retrieved at dawn. By 3 to 4 months of age, the juveniles follow their mothers and occasionally ride on their backs. By 6 months juveniles are independent and sleep alone (Charles-Dominique 1977). Walker's (1968) hand-reared baby clung to its surrogate mother and only started making short trips on its own after 2 months. Major excursions began after 6 months, by which time play behaviour was well established.

A 'clicking' vocalization was observed soon after birth, and growls and active defence behaviour appeared at 130 days in one mother-reared infant born in captivity (Grand *et al.* 1964).

Preventative medicine

No specific data are available.

Disease and mortality

No specific data are available.

Indications for hand-rearing

Charles-Dominique (1977) reported that mothers make no effort to retrieve weak newborn babies if they fall to the ground. There are no specific data for the species.

Reintegration

No specific information is available.

References

Charles-Dominique, P. (1974). Ecology and feeding behaviour of five sympatric lorisids in Gabon. In *Prosimian biology* (ed. R.D. Martin, G.A. Doyle, and A.C. Walker), pp. 131–50. Duckworth, London.

Charles-Dominique, P. (1977). *Ecology and behaviour of nocturnal primates*. Duckworth, London.

Charles-Dominique, P. (1979). Ecology and feeding behaviour of five sympatric lorisids in Gabon. In *Primate ecology: problem oriented field studies* (ed. R.W. Sussman), pp. 5–22. Wiley, New York.

Charles-Dominique, P. (1984). Bush-babies, lorises and pottos.

In *The encyclopaedia of mammals*, Vol. 1 (ed. D. MacDonald), pp. 332–7. Allen & Unwin, London.

Charles-Dominique, P. and Bearder, S.K. (1979). Field studies of lorisid behaviour: methodological aspects. In *The study of prosimian behaviour* (ed. G.A. Doyle and R.D. Martin), pp. 567–629. Academic Press, New York.

Cowgill, U.M. (1969). Some observations on the prosimian *Perodicticus potto*. *Folia Primatologica*, **2**, 144–50.

Cowgill, U.M. (1974). Co-operative behaviour in *Perodicticus potto*. In *Prosimian biology* (ed. R.D. Martin, G.A. Doyle, and A.C. Walker), pp. 261–72. Duckworth, London.

Crandall, L.S. (1964). *Management of wild mammals in captivity*. University of Chicago Press.

Grand, T., Duro, E., and Montagna, W. (1964). Potto born in captivity. *Science*, **145**, 663.

Ioannou, J.M. (1966). The oestrous cycle of the Potto. *Journal of Reproduction and Fertility* **11**, 455–57.

Jewell, P.A. and Oates, J.F. (1969). Breeding activity in prosimians and small rodents in West Africa. *Journal of Reproduction and Fertility*, **6**, (suppl.), 23–38.

Lee, P.C., Thornback, J., and Bennett, E.L. (1988). *Threatened primates of Africa*. IUCN, Gland, Switzerland.

Manley, G.H. (1966). Reproduction in lorisoid primates. *Symposia of the Zoological Society of London*, **15**, 493–509.

Pollock, J.I. (1986). The management of prosimians in captivity for conservation and research. In *Primates. The road to self-sustaining populations* (ed. K. Benirschke), pp. 269–88. Springer-Verlag, New York.

Schwartz, J.H. (1975). Development and eruption of the premolar region of prosimians and its bearing on their *Lemur biology* evolution. In (ed. I. Tattersall and R.W. Sussman), pp. 41–64. Plenum Press, New York.

Walker, A. (1968). A note on hand-rearing a potto *Perodicticus potto*. *International Zoo Yearbook*, **8**, 110–11.

Wolfheim, J.C. (1983). *Primates of the world*, pp. 44–8. University of Washington Press, Seattle.

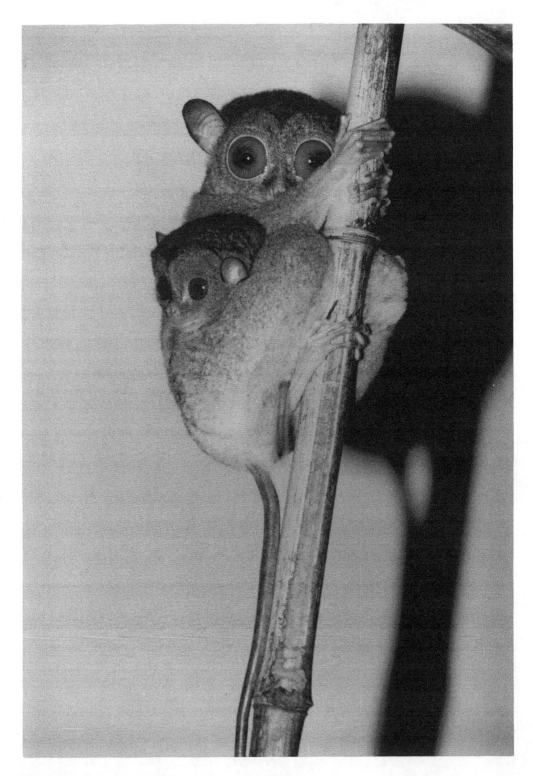
Horsfield's or western tarsier

5 Horsfield's or western tarsier

Species
Horsfield's or western tarsier   *Tarsius bancanus*

ISIS No.   1406005001001

Status, subspecies, and distribution
Horsfield's tarsier is found in northern and western parts of Borneo, eastern Sumatra, parts of Java, and on the islands of Bangka and Belitung (Mackinnon 1986; Wolfheim 1983). It has been classified as endangered but is not rare in North Borneo (Wolfheim 1983). MacKinnon (1986) estimated the total population at about 10 million, and it is not currently classified as threatened (IUCN, 1990).

It inhabits primary and secondary forest and prefers forest edges (Wolfheim 1983; Fogden 1974). Its habitat has been greatly reduced in recent times (MacKinnon 1986).

The ISIS records for 1984 included 12 individuals of the Tarsiidae family of which none were captive-born (Flesness 1986). Pollock (1986) estimated the world captive population of *T. bancanus* to be nine in 1985. It appears that the species was first bred in captivity in 1983 in Washington (Jones 1986).

Sex ratio
No specific information is available: the sex ratio among adults is 1:1 (Fogden 1974).

Social structure
Horsfield's tarsiers live as solitary individuals or in pairs (Fogden 1974), but males are necessarily monogamous. Juveniles stay with their mothers until these become pregnant again (Niemitz 1979, 1984*b*).

Breeding age
In *T. spectrum*, sexual maturity is thought to be reached during the second year (Niemitz 1984*b*), and this probably also applies to *T. bancanus*.

Longevity
Tarsius bancanus can survive up to 12 years in captivity (Crandall 1964; Niemitz 1984*b*).

Seasonality
Little information is available. However, the species is thought to have a sharply defined breeding season in the wild, with births occurring in January and February according to the food supplies (Fogden 1974).

Gestation
Gestation is thought to be about 180 days (Niemitz 1984*c*; Kohn 1986).

Pregnancy diagnosis
No specific data are available.

Birth
No specific data are available.

Litter size
Tarsiers produce single offspring (Fogden 1974).

Adult weight
Adults weigh between 100 and 140 g (Fogden 1974; Maier 1984; Stephan 1984) with an average weight of between 120 and 125 g.

Neonate weight

It is thought that newborn *T. bancanus* weigh about 24 g (Fogden 1974).

Adult diet

Tarsius bancanus is unusual amongst primates in being strictly carnivorous (Pollock 1986). It feeds on insects and vertebrates but not vegetable matter. Its diet consists mainly of beetles, grasshoppers, ants, cockroaches, butterflies, and other insects, but it may also take birds, bats, snakes, and lizards (Fogden 1974; Niemitz 1979, 1984*a*). In captivity, it does not take fruits, vegetables or bread (Niemitz 1984*a*), but may be fed grasshoppers, mealworms, small crabs, lizards, bird flesh (including chicken), beef, liver, mice, small fish and shrimps, horse flesh, and whole milk (Crandall 1964). It would be wise to add a vitamin and mineral supplement to this diet. Kohn (1986) maintained them exclusively on live crickets that had been fed a calcium-enriched diet.

Tarsius bancanus lacks the enzyme L-gluconolactone oxidase and is therefore unable to synthesize vitamin C. It is thought to obtain this vitamin from its invertebrate diet (Pollock 1986).

Adult energy requirements

It is estimated that tarsiers in the wild eat approximately one-tenth of their body weight per night (Niemitz 1979). The energy requirement of adults was estimated to be 14 kcal/d (Niemitz 1984*a*). This is equal to 68 kcal/d per $kg^{0.75}$. Until this low energy requirement is confirmed it would be wise to provide more.

Growth

Figure 5.1 shows the growth of a wild-caught infant which weighed 38 g at capture. This infant reached a weight of 55 g after 38 days when it died from pneumonia (Niemitz 1974). The average growth rate was therefore about 0.5 g/d.

Fogden (1974) considered that tarsiers reach adult weight at 15 to 18 months of age, but the growth rate of the baby hand-reared by Niemitz indicates that adult weight would be reached con-

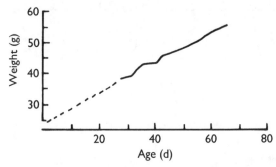

Fig. 5.1. Growth rate of one hand-reared tarsier from 38 to 55 g. Its age at 38 g was unknown but it is estimated here as 28 days. The broken line is an estimate. From Niemitz (1974).

siderably earlier than this, perhaps by 3 to 6 months of age.

Milk and milk intake

No data are available on the composition of tarsier milk. Niemitz (1974) used the human milk formula Lactogen for hand-rearing tarsiers, and provided a 10 per cent more dilute solution of this milk than is recommended for humans.

Niemitz (1974) recorded a maximum food intake of approximately 7 g/day, including 2 g of liquid formula, before the 55 g infant died. It is likely that this intake represented about 14 kcal/d, and thus that per $kg^{0.75}$ daily intake was about 125 kcal/d. This is a small amount in comparison with most other primates.

Lactation and weaning

Both pectoral and inguinal glands produce milk in the lactating female but remain small in size throughout lactation. Milk intake is thought to be 4 to 6 ml daily in the first few days of life and to increase to 6 to 8 ml daily after 2 weeks. Infants are thought to be weaned at 2 months of age (Kohn 1986). However, in the wild, an infant of 70 to 100 days of age, although able to catch prey, was observed to suckle its mother for 20 minutes (Niemitz 1984*a*). It is thought that the weaning process is gradual. Predatory behaviour appears at about 4.5 weeks, but infants will take some solid food at the end of the first week of life (Niemitz 1979). The infant is probably fed its first solid foods by its mother, and solid foods form about 30 per cent of the diet from days 9 to 20.

Feeding

Two infants obtained from the wild weighing 28.5 g and 38 g were fed on Lactogen. The formula was mixed with 10 per cent more water than prescribed for human infants (Niemitz 1974). The milk was fed *ad libitum* and the infant weighing 28.5 g took between 0.1 and 1.5 ml per feed on the first day. Nine feeds were offered and intake appeared lower during the day than at night. The same infant on the third day (weighing 32 g) first accepted a mealworm and thereafter, solids gradually formed a substantial proportion of the diet. The second infant (weighing 38 g) was fed the same formula as above, and started taking solid foods two days later when those were offered. Predatory behaviour (insect catching) was observed a week later.

Accommodation

No specific data are available.

Infant management notes

No specific data are available.

Physical development

At birth, the eyes are open and the body is covered with fur (Niemitz 1979). The development is precocious and babies can perch on vertical branches from birth (Kohn 1986).

A 28 g infant was found with its ears half unfolded; they completely unfolded four days later (Niemitz 1974).

Behavioural development

Babies can crawl around by the second day after birth. The mothers transport them by mouth whenever necessary. They appear to have a well-developed olfactory sense at birth. During the first week quadrupedal locomotion and some self-grooming are observed. After the first week, infants start to take solids, and become more active. By 8 to 10 days of age they no longer sleep at night. Vision improves at this time and the infant can focus on an object 30 cm away. At this stage threatening behaviour also appears. At 2 weeks of age, infants take their first leaps, and then this form of locomotion is preferred to climbing. Predatory behaviour appears at approximately 4.5 weeks and play behaviour a week later (Niemitz 1979). Young tarsiers in captivity first learn to catch their prey on the ground (Kohn 1986).

Diseases and preventative medicine

No specific data are available.

Indications for hand-rearing

No specific data are available.

Reintegration

No specific data are available.

References

Crandall, L.S. (1964). *Management of wild mammals in captivity.* University of Chicago Press.
Flesness, N.R. (1986). Captive status and genetic considerations. In *Primates. The road to self-sustaining populations* (ed. K. Benirschke), pp. 845–56. Springer-Verlag, New York.
Fogden, M.P.L. (1974). A preliminary field study of the western tarsier, *Tarsius bancanus*, Horsefield. In *Prosimian biology* (ed. R.D. Martin, G.A. Doyle, and A.C. Walker), pp. 151–65. Duckworth, London.
IUCN (1990) 1990 IUCN red list of threatened animals. IUCN, Gland, Switzerland.
Jones, M.L. (1986). Successes and failures of captive breeding. In *Primates. The road to self-sustaining populations* (ed. K. Benirschke), pp. 251–60. Springer-Verlag, New York.
Kohn, F. (1986). Ghost monkey. *Zoogoer* (Newsletter of the National Zoo, Washington), **15**, 10–24.
MacKinnon, K. (1986). The conservation status of non-human primates in Indonesia. In *Primates. The road to self-sustaining populations* (ed. K. Benirschke), pp. 99–126. Springer-Verlag, New York.
Maier, W. (1984). Functional morphology of the dentition of the Tarsiidae. In *Biology of tarsiers* (ed. C. Niemitz), pp. 45–58. Fischer Verlag, Stuttgart.
Niemitz, C. (1974). A contribution to the post-natal behavioural development of *Tarsius bancanus*, Horsfield 1821, studied in two cases. *Folia Primatologica*, **21**, 250–76.
Niemitz, C. (1979). Outline of the behaviour of *Tarsius bancanus*. In *The study of prosimian behaviour* (ed. G.A. Doyle and R.D. Martin), pp. 631–60. Academic Press, New York.
Niemitz, C. (1984*a*). Synecological relationships and feeding

behaviour of the genus Tarsius. In *Biology of tarsiers* (ed. C. Niemitz), pp. 59–75. Fischer Verlag, Stuttgart.

Niemitz, C. (1984b). An investigation and review of the territorial behaviour and social organisation of the genus Tarsius. In *Biology of tarsiers*, (ed. C. Niemitz), pp. 117–27. Fischer Verlag, Stuttgart.

Niemitz, C. (1984c). Tarsiers. In *The encyclopaedia of mammals,* vol. 1 (ed. D. MacDonald), pp. 338–9. Allen & Unwin, London.

Pollock, J.I. (1986). The management of prosimians in captivity for conservation and research. In *Primates. The road to self-sustaining populations* (ed. K. Benirschke), pp. 269–88. Springer-Verlag, New York.

Stephan, H. (1984). Morphology of the brain in Tarsius. In *Biology of tarsiers* (ed. C. Niemitz), pp. 319–44. Fischer Verlag, Stuttgart.

Wolfheim, J.H. (1983). *Primates of the world*, pp. 51–3. University of Washington Press, Seattle.

Common marmoset

6 Common marmoset

Species
The common marmoset *Callithrix jacchus*

ISIS No.   1406007002007001

Status, subspecies, and distribution
The common marmoset occurs in the eastern and coastal forests of Equatorial Brazil. It has also been introduced into forests as far south as Rio de Janeiro. It is not threatened at present (Mittermeier *et al.* 1986, IUCN 1990).

The species breeds well in captivity, is kept in many zoos, and is widely used in biomedical research. Johnsen and Whitehair (1986) estimated the breeding population in the principal research institutions in the United States to be 443. The number kept for research in the United Kingdom may well exceed this. Common marmosets are also becoming more widely kept as pets.

Sex ratio
Of 337 births, infants born in the Aberystwyth colony, 179 (53 per cent) were males and 158 (47 per cent) were females (Poole and Evans 1982). The ratio did not differ significantly from 1:1.

Social structure
Common marmosets live in extended family groups comprising the parents and their mature, subadult, juvenile, and infant offspring (Sommer 1984). Groups may consist of up to about 20 individuals. Infants are carried by their parents and by older siblings (Epple 1978). The dominant female suppresses oestrus in other mature females in the group (Abbott and Hearn 1978).

Breeding age
Females commence ovarian cyclicity from about 400 days of age but full sexual maturity is not reached until 20 to 24 months of age. Conceptions occurring before this are particularly likely to result in abortion or neonatal loss (Hearn 1982).

Longevity
There are reports of females breeding up to 16 years of age (Hearn 1987), and the maximum lifespan may be about 20 years.

Seasonality
Common marmosets breed throughout the year (Hearn 1982), and there is no evidence of a seasonal influence on reproductive rate (Brand 1980). Ovulation occurs 5 to 17 days after parturition (Hearn 1982), and interbirth interval averaged 154 days in one colony (Poole and Evans 1982). Common marmosets can regularly produce two litters a year. The duration of the oestrous cycle is about 28 days (Hearn 1982).

Gestation
The average gestation period is 144 days (Hearn 1982).

Pregnancy diagnosis
Pregnancy can be diagnosed by palpation of the lower abdomen from about 30 days after conception (Hearn 1982). The abdomen becomes visibly distended in late pregnancy and this is most clearly apparent when a female clings vertically to the front mesh of a cage.

Birth
Births normally occur at night, typically between 21.00 and 23.00 h (Stevenson 1976*b*). If females are observed in parturition during the day it is probable that they are in difficulties. They should be observed closely and if birth does not occur within an hour, they should be caught for exami-

nation and assistance if necessary. Caesarean section by a ventral midline incision is relatively straightforward.

The average duration of labour is about 1 hour, and there are intervals of 2 to 20 minutes between births of the infants (Stevenson, 1976b: Rothe 1978). The placentae, which are fused, are normally passed 20 to 30 minutes after birth (Stevenson 1976b) and are eaten.

Litter size

A tendency for the litter size to increase with time has been observed in several colonies (Hiddleston 1978; Lunn and Hearn 1978; Poole and Evans 1982). For example, Hearn and Burden (1979) found that the incidence of triplets in their colony increased from 12 per cent in 1974 to 43 per cent in 1977. Hiddleston (cited by Bruhin 1979) also reported a high (41 per cent) incidence of triplet births. The explanation for this is not clear, but it may reflect improvements to management, especially nutrition. Of 129 litters born at the Aberystwyth colony, 55 per cent were triplets, 35 per cent twins, 8 per cent quadruplets, and 2 per cent singletons (Poole and Evans 1982). Data for three other colonies reviewed by Poole and Evans (1982) showed that there was variation between colonies in average litter size but, in these others, the most common litter size was two. Forty-five out of 66 parturitions (68 per cent) reported by Tardif et al. (1984) were of twins.

Adult weight

There is little evidence for sexual dimorphism in adult weight but adult weights differ between colonies and may be influenced by the plane of nutrition during growth. Wild caught males averaged 311 g and wild-caught females averaged 291 g (Cooper, unpublished, cited by Hershkovitz 1977). Poole and Evans (1982) reported that adult males in their colony weighed 386–493 g, and females 382–600 g; but Hearn (1982) reported an average adult weight of about 310 g. Wolfe et al. (1972) observed that body weights at 2 years of age were about 10 per cent greater in hand-reared than mother-reared offspring.

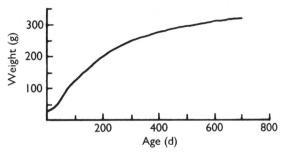

Fig. 6.1. Average growth curve of mother-reared common marmosets to maturity. From Hearn (1982).

Neonate weight

The mean neonate weight was found by Turton et al. (1978b) to be 29.3 g. This is similar to that apparent from the growth curve provided by Hearn (1982). Hampton et al. (1971) found that twins had a greater average birth weight (30.3 g) than triplets (23.6 g).

Adult diet

Common marmosets are frequently fed on primate pellets supplemented with fresh fruit, and sources of animal protein, such as cheese, hard-boiled eggs, crickets or mealworms. Care should be taken not to provide so much fruit that the intake of the other items of the diet is compromised. The pellet should include vitamin D_3. Nutritional bone disease occurs in animals whose diet lacks adequate vitamin D_3, but care should also to be taken to avoid excessive intake.

Adult energy requirements

The energy requirement for maintenance is likely to be close to 120 kcal/d per $kg^{0.75}$, i.e., about 55 to 65 kcal/d per adult animal.

Growth

The observations of Hearn (1982) indicated an average growth rate of about 1 g/d for the first 100 days after birth, and thereafter a gradual decline in growth rate until adult weight is reached at about 2 years of age (Fig. 6.1). Turton et al. (1978b) provided more detailed graphs of the growth of mother-reared babies to 35 days of

Fig. 6.2. Average neonatal growth rate of collaboratively reared common marmosets. From Hearn and Burden (1979).

age which indicated that, in these early stages of growth, the average gain was about 0.6 g/d. Records for hand-reared babies provided by these authors show several failed to gain weight during the first 5 to 10 days, and some lost a few grams in this period (see also Fig. 6.2).

The weight gain and growth of various internal organs of the fetus has been described by Chambers and Hearn (1985).

Milk and milk intake

Details of the composition milk collected by manual expression (without using sedation or oxytocin) from common marmosets at various stages of lactation have been documented by Turton et al. (1978a). The milk contained 7.7 per cent fat, 3.6 per cent protein, and 7.5 per cent lactose, and had an energy density of 1.14 kcal/ml. If we assume an ash content of 0.3 per cent (as, for example, in the squirrel monkey), then the total solids must be about 19.1 per cent. The proportions of fat, protein, and lactose in the dry matter must be about 0.40, 0.19, and 0.39 respectively. These results indicate that the milk is more concentrated than human milk, and contains higher proportions of fat and protein.

Turton et al. (1978a) found the osmotic pressure of the milk to be 354 mmol/kg water, and measured the concentration of some minerals: sodium 21.4, potassium 54.3, calcium 92.2, phosphorus 22.8, magnesium 5.0, and chloride 52.2 mg per 100 ml whole-milk. A detailed analysis of the fatty acid composition of the milk was also undertaken by Turton et al. (1978a), who pointed out, among other things, that marmoset milk contains considerably higher proportions of long chain polyunsaturated fatty acids than human milk and human milk replacers.

Human milk replacers, such as SMA, have been used successfully to rear common marmosets (Ingram 1975; Pook 1976), but babies reared by Stevenson (1976a) failed to grow on this diet. These formulae have the advantage of being readily available and of reliable quality. However, their composition differs from that of marmoset milk and is not therefore ideal. Turton et al. (1978a) suggested that feeding unsupplemented human milk replacers would be unlikely to promote satisfactory growth and development due to the different fatty acid, protein, and energy levels in marmoset milk. They suggested that formulae based on human milk replacers should be supplemented to mimic marmoset fat, protein, and carbohydrate levels, and that the addition of 450–500 mg of cod liver oil per 100 ml would assist in providing long chain polyunsaturated fatty acids.

Turton et al. (1978b) devised a formula with a composition of 7.6 per cent fat, 3.3 per cent protein, 5.1 per cent lactose, and 1.02 kcal/ml, by mixing 76 ml Goldcap SMA Ready-to-Feed Milk, 24 ml of pure Dairy Sterilised Cream (Nestlé), and 2 g calcium caseinate (Casilan). Some babies were reared successfully using this formula. Stevenson (1976a) used a formula based on the same ingredients (13.2 g SMA powder, 5.4 g calcium caseinate, and 84 ml water) but with a composition of 3.7 per cent fat, 6.8 per cent protein, and 5.7 per cent lactose. Hearn and Burden (1979) also used formulae based on SMA and Casilan but with added glucose, and Abidec and Cytacon supplements. For babies aged 2–10 days, they mixed, presumably in equal proportions, a 2 per cent glucose solution with SMA and Casilan solutions (each made up at 30 g per 100 ml) and they added 1 Abidec drop and 5 drops of Cytacon to the mixture. For younger babies, the mixture comprised 10 per cent glucose mixed with 20 per cent SMA and 20 per cent Casilan solutions, and for those aged 10–30 days, they used 2 per cent glucose mixed with solutions of SMA and Casilan both made up at 50 g per 100 ml water.

Wolfe et al. (1972) reported the successful use of a formula composed of 1 tablespoonful of SMA powder and 1 teaspoonful of Sustagen added to 56 ml of boiled water for a variety of *Saguinus* species, and the use of this formula for Callithrix has been described by others (Ogden 1979; Anderson 1986).

Cicmanec et al. (1979) used 15 g each of SMA and Sustagen per 60 ml boiled water. The dry matter content of this formula would be close to 33 per cent, considerably greater than that of natural marmoset milk. As these authors pointed out, there is a danger of dehydration with fatal consequences if too high a concentration is used.

Primilac contains higher fat and protein levels than human milk replacers and may be a suitable off-the-shelf product for this species. There are insufficient data on growth and survival rates for critical comparison of performance on the various milk formulae that have been used.

The literature indicates that milk intake is typically about 2 to 5 ml per day initially, rising to about 7 to 12 ml per day by 10 days of age, and reaching a peak of about 20 ml per day at around 3 weeks of age (Pook 1976; Stevenson 1976a; Cicmanec et al. 1979; Hearn and Burden 1979; Ogden 1979). These intakes have been achieved by feeding every 2 to 3 hours (in some cases with a break between midnight and 8.00 h), with meal sizes increasing from about 0.3 to 0.5 ml in the first day or two, to about 2 ml per feed by three weeks of age. These figures should only be used as a rough guide because there is considerable variation between individuals.

None of the accounts of hand-rearing of this species provide sufficient detail to rigorously assess energy intake rates in the early stages of growth. However, as formulae typically have energy densities of about 0.7 to 1 kcal/ml, in the first few days of life energy intake of hand-reared babies appears to have been around 3 to 6 kcal/d (Fig. 6.3), which is, as neonate weight is about 25 to 30 g, in the order of 50 to 80 kcal/d per metabolic weight (Fig. 6.4). The energy intake of Hearn and Burden's 'collaboratively' reared babies (they assisted in the rearing of triplets by

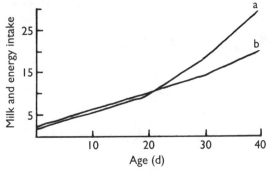

Fig. 6.3. Approximate daily (a) milk, ml/d, and (b) energy intake, kcal/d, provided as a supplement to collaboratively reared common marmosets in relation to age. From Hearn and Burden (1979).

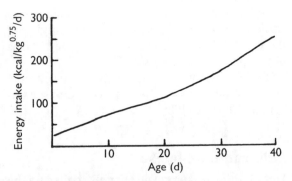

Fig. 6.4. Approximate daily energy per unit metabolic weight provided as supplementary feeds to collaboratively reared common marmosets in relation to age. From Hearn and Burden (1979).

taking one, in turns, away each day for supplementary feeding) is shown in Fig. 6.3. The milk-energy intake of these babies increased to about 170 kcal/d per metabolic weight at 30 days of age (Fig. 6.4).

Lactation and weaning

Newborn common marmosets commonly find the teat and start sucking within 2 to 20 minutes of birth (Rothe 1978). Lactation may continue for up to 90 to 100 days post-partum (Lunn and Hearn 1978), but infants begin to take solid food at about 21 days of age (Pook 1976) and can survive without their mothers from about 40 days of age (Hearn 1987). Epple (1970) considered that weaning took place between 52 and 85 days.

The process of weaning can be started at 14 to 21 days of age by adding human baby cereal products, such as Farex, to the milk. Turton et al. (1978b) added a tinned infant beef and liver soup to the diet of 30 day-old babies, small quantities of egg at 60 days of age, and fruit at about 90 days of age. Others (for example, Pook 1976; Stevenson 1976a; Hearn and Burden 1979; Ogden 1979) have offered solid foods, such as chopped fruit and primate pellets, to babies of about 3 weeks of age. There is little to be lost by making small quantities of the adult diet available for the baby to eat if it wishes to at this stage.

Feeding

Hearn and Burden (1979) fed their baby marmosets by gently expressing milk from a 1 ml syringe and allowing the babies to lap at the meniscus. Others have used small latex nipples attached to 2–3 ml bottles or syringes (for example, Pook 1976; Stevenson 1976a; Ogden 1979). Ogden's (1979) method for making nipples is described in the section for cotton-top tamarins (Chapter 7). Baby common marmosets are tiny and great care must be taken not to choke them with too fast a flow of milk. Inhalation of milk may lead to pneumonia.

Ogden (1979), in addition to hand-feeding provided milk formula in drinkers, attached to the cage walls, from which the babies could lap. The sippers of these were guarded by rubber mouth pieces to prevent trauma. He reported that they began self-feeding after 2 weeks of age, and were totally self-feeding by 4 to 6 weeks of age.

Newborn common marmosets may require night feeds during the first two weeks (Turton et al. 1978b) although Ingram (1975) did not consider this necessary. After that, feeding every two to four hours between 8.00 h and 12.00 h is adequate. The frequency of feeding can be gradually reduced after 14 days of age if progress is satisfactory.

Accommodation

The notes on accommodation for infant cotton-top tamarins are equally relevant to the common marmoset. Hearn and Burden (1979) used incubators at 25–30 °C and a relative humidity of 80 per cent. Initially, a higher ambient temperature of 30 to 35 °C may be preferable, but great care must be taken not to exceed this upper limit. They provided soft toys for the babies to cling to as surrogate mothers. Ogden (1979) made surrogates by packing 5 cm diameter stockinette with cotton wool and suspending these, at an angle of about 60 degrees (from the horizontal) from the roof of the incubator by elastic bands. The advantage of this type of surrogate is that it is disposable when soiled (see the notes on the cotton-top tamarin and squirrel monkey for details of more sophisticated surrogate mothers, Chapters 7 and 10).

Ambient temperature should be gradually reduced to room temperature (ideally around 25 °C).

Infant management notes

After each feed, gentle massage of the perineum using a small cotton wool swab dipped in baby oil helps to stimulate urination and defaecation.

Infants should be weighed each morning before feeding to monitor growth and assist in the early detection of disease or suboptimal management.

Babies should be housed so that they can see and hear other marmosets from the first few days of life (Ingram 1975). Hand-rearing in isolation from others of the species must be avoided, as in all primate species. One baby reared in isolation from 14 days of age showed self-biting behaviour up to 21 months of age (Berkson et al. 1966).

Physical development

At birth the babies are covered with short grey hair and their eyes are usually open, although they may remain closed for 3 or 4 days after parturition. At 10–12 weeks the colour of the body is like that of the adults, and by 6–7 months the typical adult face pattern and white aural corollae are apparent (Epple 1970).

The upper and lower medial incisor teeth erupt at 2–8 days, followed by the lateral incisors at 5–11 days (the lower ones first). The lower canines erupt at 14–20 days and the upper ones at 23–27 days. There are no premolars, and the molars (first, second, and third) erupt sequentially at 1–3 day intervals after the canines of the same jaw (Swindler 1976; Winter 1978).

The first permanent teeth to erupt are the first molars at 106 to 150 days, followed by: second molars, 164–228 days; medial incisors, 188–195 days; third premolars, 210–250 days; second premolars and lateral incisors, 230–300 days; first premolars, 260–310 days; and canines, 280–320 days (Winter 1978).

Behavioural development

Babies are continuously carried by their parents and others in their family groups for the first 3 weeks, after which they begin to make their first independent forays. They may continue to ride on their parents' backs occasionally until at least 3 months of age, and cling to their backs at night until 12–17 weeks of age (Epple 1970). Self-grooming behaviour was first noted at 13 days of age by Pook (1976).

Disease and mortality

Neonatal mortality is often reported to be high in captivity (Sainsbury 1989). Although triplets and quadruplets are quite often born in captivity, they are very rarely all reared, and usually one weakens and dies within the first few days, although there are reports of triplets being successfully mother-reared. The failure to rear three may be due to a limit to the mother's milk production or competition for the two teats. It is interesting that golden lion tamarins, in contrast, often successfully rear triplets (Kleiman et al. 1982).

Poole and Evans (1982) compared pre- and postnatal mortality rates between colonies. Prenatal mortality has been estimated at between 4 per cent (Hiddleston 1976) and 30 per cent (Hampton et al. 1978), and Poole and Evans (1982) considered that it accounted for 11 per cent of pregnancies in their own colony. First pregnancies, particularly in very young females, may be particularly likely to be aborted (Hearn 1982).

Still births have been common in colony-born babies. For example, 28 out of 100 babies were born dead at one center (Johnson et al. 1986), 12 out of 136 (9 per cent) at another (Tardif et al. 1984), and 29 out of 275 (10.5 per cent) in the colony described by Lunn and Hearn (1978).

Of 124 liveborn babies at the colony described by Tardif et al. (1984), 31 (25 per cent) died prior to 3 months of age, and mortality was higher among triplets [11 (37 per cent) of 30], than amongst twins [20 (23 per cent) of 88], and singletons (0 of 6) (Tardif et al. 1984). Pre-weaning mortality rates have been published for other colonies, for example 58 per cent (Hampton et al. 1978), 27 per cent (Hiddleston 1976), and 36 per cent (Poole and Evans 1982). At another center, the mortality prior to 30 days of age was 24 per cent, and prior to 2 years of age was 50 per cent (Johnson et al. 1986).

There are few published accounts of the mortality rate in hand-reared babies. Of 344 babies, some normal and some abandoned, of a variety of species of Callitrichidae hand-reared by Ogden (1979), 109 (32 per cent) died prior to 30 days of age. He recorded a similar mortality prior to 30 days of age (37 per cent) in mother-reared babies in his colony.

In the early neonatal period many deaths are due to starvation. This usually occurs in the weakest baby of a triplet litter. As in other primates, the previous social environment of the parents is also an important factor. Mortality is higher among babies born to mothers that have not previously gained experience assisting in carrying younger siblings in their parental groups (Phillips 1976). Mothers that have had an abnormal social development are more likely to abandon or traumatize their babies, particularly those of their first litter. High neonatal losses are common amongst females that are paired for breeding when too young, before about 15 months of age (Hearn 1987).

There have been no thorough analyses of causes of death in the later stages of growth in colony-born, mother-reared marmosets as far as we are aware, although it is our impression that pneumonia is quite common. Bite injuries to tails and limbs inflicted by other cage mates are also quite common (Cicmanec et al. 1979).

Among hand-reared babies, common problems are aspiration pneumonia and enteritis. Aspiration of milk can occur if the milk flow is too high, or following regurgitation after ingesting too great a volume (Cicmanec et al. 1979). Overfeeding can also lead to a nutritional diarrhoea and enteritis. Cicmanec et al. (1979) also found bacterial pneumonias and oral candidiasis to be quite common.

The discussion of the effects of cage size and environmental complexity on infant mortality of cotton-top tamarins is also pertinent to this species.

Preventative medicine

Much of the infant mortality that occurs in colony-born common marmosets has a management component. The high incidence of triplet litters may be related to the captive environment, but the causal mechanism is not known. Breeding pairs should be kept in large cages so that offspring can remain in the family group long enough to gain experience with at least one litter of younger siblings before being removed for pairing. Disturbance should be minimized and high standards of hygiene and husbandry should be maintained.

Ogden (1979) added ampicillin (1 drop of a 100 mg/ml preparation) to each feed until the babies were 10 days old. His survival rates were good, but disadvantages to the prophylactic use of antibiotics in healthy babies can be envisaged. First, these may interfere with the establishment of a normal bacterial flora and facilitate fungal infections (for example, *Candida*), and secondly, the common pathogens are likely to become resistant.

Common marmosets and other Callitrichidae are susceptible to *Herpesvirus hominis* — the virus that causes 'cold sores' in man, and infection can kill them (Daniel et al. 1978). If they have these lesions or show signs of other infectious diseases, they should be isolated from personnel.

The New World monkeys need a dietary source of vitamin D_3, without it, normal calcification of the bones fails. Human milk replacers and Primilac contain adequate levels of this vitamin, as do reputable pelleted primate diets. Additional supplements of vitamin D_3 are, however, often provided and may be a useful safeguard at the time of weaning.

Indications for hand-rearing

Babies that have been abandoned and/or traumatized should either be humanely killed or taken for cross-fostering or hand-rearing. When the litter size is greater than two, it is unlikely that more than two babies will survive. One can be taken for hand-rearing or cross-fostering, or supplementary feeding can be provided for the litter.

Hearn and Burden (1979) removed each baby from a triplet litter in turn for a day during which it was kept on a soft-toy surrogate mother in an incubator and hand-fed. This regime was continued until the litter was 40 days old. These authors found that there were no rejections as a result of disturbance in removing the babies or on returning them.

If there is an available lactating female who has recently lost her own litter, or who only has a single baby, it may be possible to cross-foster an abandoned baby to her. However, females do not always accept other babies and may kill them. It may help to rub the baby over the body and vulval scent glands, of the foster mother, and then to allow it to cling to the mother's own baby in the nest box for 30 minutes before allowing the mother in (Hearn 1987).

An alternative is to foster the baby with an experienced pair that have no young of their own, and remove it regularly for hand-feeding. This has been found to be successful by several workers (Stevenson and Sutcliffe, 1978; Harris, cited by Hearn 1987).

Reintegration

Two babies reared by Poole and Evans (1982) together but out of direct contact with older marmosets, showed behavioural abnormalities as adults and failed to breed. Others hand-raised by these authors were placed in a cage with older hand-reared marmosets from 2 weeks of age, for increasing periods (initially 30 minutes per day) each day. Babies reared in this way all later proved capable of breeding. This system was also advocated by Stevenson (1976a), who reported that one infant, at 34 days of age, took to being carried by the 5-month juveniles in whose cage she was regularly placed.

Hand-reared baby marmosets can, if introduced to older marmosets at an early age, breed successfully as adults. The notes on reintroduction of squirrel monkeys (Chapter 10) and cotton-top tamarins (Chapter 7) are of relevance to this species.

References

Abbott, D.H. and Hearn, J.P. (1978). Physical, hormonal and behavioural aspects of sexual development in the marmoset monkey, *Callithrix jacchus*. *Journal of Reproduction and Fertility*, **53**, 155–66.

Anderson, J.H. (1986). Rearing and intensive care of neonatal and infant nonhuman primates. In *Primates. The road to self-sustaining populations* (ed. K. Benirschke), pp. 747–62. Springer-Verlag, New York.

Berkson, G., Goodrich, J., and Kraft, I. (1966). Abnormal stereotyped movements of marmosets. *Perceptual and Motor Skills*, **23**, 491–8.

Brand, H.M. (1980). Influence of season on birth distribution on marmosets and tamarins. *Laboratory Animals*, **14**, 301–2.

Bruhin, von H. (1979). *Callithrix jacchus*, marmoset. Ein neues versuchsmodell. *Zeitschrift für Versuchstierkunde*, **21**, 209–21.

Chambers, P.L. and Hearn, J.P. (1985). Embryonic, foetal and placental development in the common marmoset monkey (*Callithrix jacchus*). *Journal of Zoology (London)*, **207A**, 545–61.

Cicmanec, J.L., Hernandez, D.M., Jenkins, S.R., Campbell, A.K., and Smith, J.A. (1979). Hand-rearing infant Callitrichids (Saguinus spp and *Callithrix jacchus*), owl monkeys (*Aotus trivirgatus*), and capuchins (*Cebus alibifrons*). In *Nursery care of nonhuman primates* (ed. G.C. Ruppenthal), pp. 307–12. Plenum Press, New York.

Daniel, M.D. et al. (1978). Prevention of fatal herpesvirus infections in owl and marmoset monkeys by vaccination. In *Recent advances in primatology*, Vol. 4. *Medicine* (ed. D.J. Chivers, and E.H.R. Ford), pp. 67–9. Academic Press, London.

Epple, G. (1970). Maintenance, breeding and development of marmoset monkeys (Callithricidae) in captivity. *Folia Primatologica*, **13**, 48–62.

Epple, G. (1978). Reproductive and social behaviour of marmosets with special reference to captive breeding. In *Primates in medicine* (ed. E. Goldsmith and J. Moor-Jankowski), pp. 50–62. Karger, Basel.

Hampton, J.K. Jr., Hampton, S.H., and Levy, B.M. (1971). Reproductive physiology and pregnancy in marmosets. In *Medical primatology* (ed. E.I. Goldsmith and J. Moor-Jankowski), pp. 527–35. Karger, Basel.

Hampton, S.H., Gross, M.J., and Hampton, J.K. Jr. (1978). A comparison of captive breeding performance and offspring survival in the family Callitrichidae. *Primates in Medicine*, **10**, 88–95.

Hearn, J.H. (1982). The reproductive physiology of the common marmoset *Callithrix jacchus* in captivity. *International Zoo Yearbook*, **22**, 138–43.

Hearn, J.H. (1987). Marmosets and tamarins. In *The UFAW handbook on the care and management of laboratory animals* (ed. T.B. Poole), pp. 568–81. Longman, London.

Hearn, J.P. and Burden, F.J. (1979). 'Collaborative' rearing of marmoset triplets. *Laboratory Animals*, **13**, 131–3.

Hershkovitz, P. (1977). *Living New World monkeys (Platyrrhini)*, Vol. 1. University of Chicago Press.

Hiddleston, W.A. (1976). Large scale production of a small laboratory primate *Callithrix jacchus*. In *The laboratory animal in the study of reproduction* (ed. T. Antikatzades, S. Erichsen, and A. Spiegel), pp. 51–7. Fisher, New York.

Hiddleston, W.A. (1978). The production of the common marmoset *Callithrix jacchus* as a laboratory animal. In *Recent advances in primatology*, Vol. 2 (ed. D.J. Chivers and W. Lane-Petter), pp. 173–81. Academic Press, New York.

Ingram, J.C. (1975). Husbandry and observation methods of a breeding colony of marmosets (*Callithrix jacchus*) for behavioural research. *Laboratory Animals*, **9**, 249–59.

IUCN (1990). *1990 IUCN red list of threatened animals*. IUCN, Gland, Switzerland.

Johnsen, D.O. and Whitehair, L.A. (1986). Research facility breeding. In *Primates. The road to self-sustaining populations* (ed. K. Benirschke), pp. 499–511. Springer-Verlag, New York.

Johnson, L.D., Petto, A.J., Boy, D.S., Sehgal, P.K., and Beland, M.E. (1986). Effect of mortality on colony-born production. In *Primates. The road to self-sustaining populations* (ed. K. Benirschke), pp. 772–9. Springer-Verlag, New York.

Kleiman, G.G., Ballou, J.D., and Evans, R.F. (1982). An analysis of recent reproductive trends in captive golden lion tamarins *Leontopithecus r. rosalia* with comments on their future demographic management. *International Zoo Yearbook*, **221**, 94–101.

Lunn, S.F. and Hearn, J.H. (1978). Breeding marmosets for medical research. In *Recent advances in primatology*, Vol. 2 (ed. D.J. Chivers and W. Lane-Petter), pp. 183–5. Academic Press, New York.

Mittermeier, R.A., Oates, J.F., Eudey, A.E., and Thornback, J. (1986). Primate conservation. In *Comparative primate biology*, Vol. 2A. *Behaviour, conservation and ecology* (ed. G. Mitchell and J. Erwin), pp. 3–72. Alan R. Liss, New York.

Ogden, J.D. (1979). Hand-rearing *Saguinus* and *Callithrix* genera of marmosets. In *Nursery care of nonhuman primates* (ed. G.C. Ruppenthal), pp. 313–19. Plenum Press, New York.

Phillips, I.R. (1976). The reproductive potential of the common cotton-eared marmoset (*Callithrix jacchus*) in captivity. *Journal of Medical Primatology*, **5**, 49–55.

Pook, A.G. (1976). Some notes on the development of hand-reared infants of four species of marmoset (Callitrichidae). *Jersey Wildlife Preservation Trust, Annual Report*, **13**, 38–46.

Poole, T.B. and Evans, R.G. (1982). Reproduction, infant survival and productivity of a colony of common marmosets (*Callithrix jacchus jacchus*). *Laboratory Animals*, **16**, 88–97.

Rothe, H. (1978). Parturition and related behaviour in *Callithrix jacchus*. In *The biology and conservation of the Callitrichidae* (ed. D.G. Kleiman), pp. 193–206. Smithsonian Institution, Washington, D.C.

Sainsbury, A.W. (1989). Reducing infant mortality in marmosets. In *Care of Young wild animals in captivity* (ed. P.W. Scott and J.K. Kirkwood), pp. 55–65. The British Veterinary Zoological Society, c/o British Veterinary Association, London.

Sommer, V. (1984). Dynamics of group structure in a family of the common marmoset *Callithrix jacchus* (Callitrichidae). In *Current primate researches* (ed. M.L. Roonwal, S.M. Mohnot, and N.S. Rathore), pp. 315–42. Jodhpur University.

Stevenson, M.F. (1976a). Maintenance and breeding of the common marmoset (*Callithrix jacchus*) with notes on hand-rearing. *International Zoo Yearbook*, **16**, 110–16.

Stevenson, M.F. (1976b). Birth and perinatal behaviour in family groups of the common marmoset (*Callithrix jacchus jacchus*) compared to other primates. *Journal of Human Evolution*, **5**, 376–81.

Stevenson, M.F. and Sutcliffe, A.G. (1978). Breeding a second generation of common marmosets *Callithrix jacchus* in captivity. *International Zoo Yearbook*, **18**, 109–14.

Swindler, D.R. (1976). *Dentition of living primates*. Academic Press, London.

Tardif, S.D., Richter, C.B., and Carson, R.L. (1984). Reproductive performance of three species of Callitrichidae. *Laboratory Animal Science*, **34**, 272–5.

Turton, J.A., Ford, D.J., Bleby, J., Hall, B.M., and Whiting, R. (1978a). Composition of the milk of the common marmoset (*Callithrix jacchus*) and milk substitutes used in hand-rearing programmes, with special reference to fatty acids. *Folia Primatologia*, **29**, 64–79.

Turton, J.A., Hobbs, K.R., Ford, D.J., Bleby, J., and Hall, B.M. (1978b). An artificial milk for hand-rearing specified pathogen free marmosets, *Callithrix jacchus,* and the growth of animals on the preparation. In *Recent advances in primatology*, Vol. 2 (ed. D.J. Chivers and W. Lane-Petter), pp. 187–93. Academic Press, London.

Winter, M. (1978). Investigation of the sequence of tooth eruption in hand-reared *Callithrix jacchus*. In *Biology and behaviour of marmosets* (ed. H. Rothe, H.-J. Wolters, and J.P. Hearn), pp. 109–24. Proceedings of the Marmoset Workshop, Göttingen, September 1977.

Wolfe, L.G., Ogden, J.D., Deinhardt, J.B., Fisher, L., and Deinhardt, F. (1972). Breeding and hand-rearing marmosets for viral oncogenesis studies. In *Breeding primates* (ed. W.I.B. Beveridge), pp. 145–57. Karger, Basel.

Cotton-top tamarin

7 Cotton-top tamarin

Species
The cotton-top tamarin   *Saguinus oedipus*

ISIS No.   1406007004018

Status, subspecies, and distribution
The cotton-top tamarin is an endangered species (IUCN, 1990). It is found in parts of Colombia and the population is thought to be between 1000 and 10 000 (Neyman 1977). Tardif (1986) estimated the captive population at 1450, with 40 per cent in zoos and 60 per cent in research colonies.

Sex ratio
The sex ratio of all the zoo and research colony animals that Tardif (1986) was able to collect records of was 1 female to 1.12 male. There is a marked tendency for more male than female births. For example, Evans (1983) reported 17 female and 22 male births, Snowdon *et al.* (1985) 35 female and 52 male births, and Scullion (personal communication) 52 female and 68 male births.

Social structure
In captivity, cotton-tops are successfully kept as family groups from which young adults are removed, ideally after gaining experience carrying young siblings, to establish new breeding pairs. This roughly mimics the situation in the wild where small groups consist of a breeding pair, their recent offspring, and one or two incomers from other families.

Breeding age
The average age of first birth or abortion in females of known age was 31 months (Kirkwood *et al.* 1985) and those authors reported that no female younger than 29 months had successfully reared young in their colony. Males probably reach sexual maturity at about 1 year of age.

Longevity
Both sexes have lived to at least 15 years (Colley 1986).

Seasonality
Births have been recorded in all months of the year but in some colonies in the northern hemisphere there have been more births in the spring months (March to July) than during the rest of the year (Gengozian *et al.* 1978; Brand 1980; Kirkwood *et al.* 1983, 1985), with a second smaller peak in November and December. Others have found no seasonal birth pattern (Snowdon *et al.* 1985). Interbirth interval tended to be shorter (180–280 days) in females that lost their young at birth compared with those that successfully reared them (Kirkwood *et al.* 1983). The mean interbirth interval in the colony at the Marmoset Research Center, Oak Ridge, Tennessee was 236 days (Tardif *et al.* 1984).

Gestation
The gestation period is about 166 days (Brand 1981).

Pregnancy diagnosis
Pregnancy can be diagnosed by palpation at about 5 weeks from conception. In the late stages of pregnancy the abdomen becomes visibly distended. Weight increases by about 2 g/d during the last 8 weeks of pregnancy (Kirkwood and Underwood 1984).

Birth
Parturition usually takes place at night. All nine births observed by Price (1988) occurred between

17.50 h and 20.40 h. The females showed restlessness for 2 hours before parturition. The placenta is usually eaten by the mother or other members of the family. Females observed in parturition during the morning are likely to be in difficulties having started labour the previous night. They should be observed closely and examined soon afterwards if no progress is being made.

Litter size

The litter size varies from 1 to 3. In 54 litters at one colony 8 (15 per cent) were singles, 33 (61 per cent) were twins and 13 (24 per cent) were triplets (Kirkwood *et al.* 1985). In this colony no triplet litters were recorded among 39 born prior to March 1982, but thereafter about a quarter of the litters were triplets (Kirkwood 1983). This increase in litter size may have been associated with a nutritional factor. Evans (1983) reported the size of 19 litters: 1 single, 2 triplets, and 16 (84 per cent) twins. At the Marmoset Research Center, Tennessee, 16 per cent of litters were singles, 70 per cent twins, and 20 per cent triplets (Tardif *et al.* 1984). At the University of Wisconsin, among 55 litters, 18 per cent were singletons, 80 per cent twins, and 2 per cent triplets (Snowdon *et al.* 1985).

Adult weight

Kirkwood (1983) found that the mean adult weights of colony-born males and females were 490 g and 481 g respectively. Wild-caught animals tended to weigh less and averaged about 450 g (Kirkwood 1983). Adult weight appears to vary from colony to colony, probably associated with differences in the diet.

Neonate weight

The average neonate weight at the Bristol University colony was 42 g and ranged from 36 to 52 g (Kirkwood unpublished). Hershkovitz (1977) lists average weights of 34 g for triplets, 36 g for twins, and 37 g for singles but these were from the early days of management of cotton-tops in captivity and birth weights are usually higher now, perhaps because of improved nutrition. From the

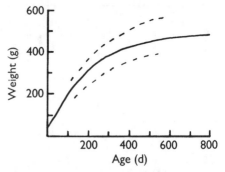

Fig. 7.1. Mean growth curve of mother-reared colony-born cotton-top tamarins. The broken lines indicate normal range. From Kirkwood (1983).

graph (Fig. 7.1) of the growth of hand-reared animals at the University of Wisconsin (Dronzek *et al.* 1986) it appears that birth weight averages about 45 g, and from that of hand-reared animals at the German Primate Centre, about 42 g (Kaumanns *et al.* 1986).

Adult diet

The literature indicates that most captive cotton-top tamarins are fed a complete pelleted ration supplemented with a variety of fruit and animal protein sources, such as cheese, egg, and mealworms (for example, Kirkwood *et al.* 1983; Snowdon *et al.* 1985; Clapp and Tardif 1985).

Adult energy requirements

In one study, daily metabolizable energy intake was found to average 76 kcal/d in adults, with a range of 50 kcal/d in an old wild-caught male weighing 451 g, to 114 kcal/d in a 2-year-old male weighing 551 g (Kirkwood and Underwood 1984). Escajadillo *et al.* (1981) reported a slightly higher mean intake. Energy requirements do not appear to increase by very much during pregnancy (perhaps by 10 per cent), but females appear to roughly double their intake during lactation (Kirkwood and Underwood 1984). The mean daily metabolizable energy intake in non-breeding adults corresponds to about 120 $kcal/kg^{0.75}$ per day.

Growth

Parent-reared cotton-tops were found by Kirkwood (1983) to reach adult weight at 600 to

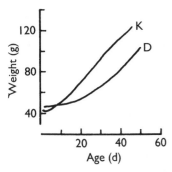

Fig. 7.2. Neonatal growth rate of hand-reared cotton-top tamarins. (D) mean of six reared by Dronzek et al. (1986). (K) mean of 11 reared by Kaumanns et al. (1986). From Dronzek et al. (1986) and Kaumanns et al. (1986).

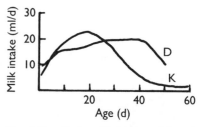

Fig. 7.3. Daily milk intake of hand-reared infant cotton top-tamarins. (D) approximate average of 6 reared by Dronzek et al. (1986). (K) approximate average of 4 males and 4 females reared by Kaumanns et al. (1986). From Dronzek et al. (1986) and Kaumanns et al. (1986).

880 days of age. Up to 200 days of age the average growth rate was about 1.8 g/d (Fig. 7.1). The six animals hand-reared by Dronzek et al. (1986) reached a mean weight of 105 g, 50 days after birth (Fig. 7.2), and the mean reached at this age by the 11 reared by Kaumanns et al. (1986) was about 120 g. The growth curve of parent-reared animals generated by Kirkwood (1983) predicts a weight of 120 g at 50 days of age. There is, however, a wide range amongst individuals and considerably slower rates of growth have been recorded in hand-reared animals (Pook 1976; Kahn 1985). No significant sex differences in growth rate have been observed in this species.

Milk and milk intake

Jenness and Sloan (1970) reported on analysis of three milk samples from *Saguinus oedipus*. These had reported a dry matter content of 13.1 per cent, and a composition of 3.1 per cent fat, 3.8 per cent protein (casein plus whey protein), 5.8 per cent lactose, and 0.4 per cent ash. The proportions of fat, protein, carbohydrate, and ash in the dry matter were therefore: 0.24, 0.29, 0.44, and 0.03 respectively.

Although the protein content of *S. oedipus* milk is higher than that of man, human milk replacers have been successfully used to rear cotton-tops, for example, SMA (Pook 1976), SMA with added Sustagen (Kahn 1985), Similac (Schmitt and Solder 1985), Similac with added Sustagen (Foster 1985; Tardif et al. 1985), and Aptamil (Kaumanns et al. 1986). Primilac might be more suitable, because of its higher protein content and for this reason it was used, with no extra vitamin supplementation, by Dronzek et al. (1986). However, no controlled comparisons have been undertaken as far as we are aware. The Similac/Sustagen mixture used by Tardif et al. (1985) contained 0.93 kcal/ml. Cicmanec et al. (1979) used a formula comprising 15 g SMA, 15 g Sustagen, and 60 ml of water. This had an energy density of 1.6 kcal/ml, and a protein content of 4.4 per cent.

The literature indicates that the milk intake on the first day is typically about 6 ml/d, rising to 10–20 ml/d by day 10 and 15–25 ml/d by day 20 (Fig. 7.3). After 30 days, the milk intake may continue to rise, but other foods begin to contribute to the daily diet. From the data of Dronzek et al. (1986) and Kaumanns et al. (1986) it appears that the metabolizable energy intake (milk intake in ml/d multiplied by its metabolizable energy density: Primilac 0.825 kcal/ml, Aptamil 0.65 kcal/ml) is quite low; 90–150 kcal/d per kg$^{0.75}$ in the early stages of growth (Fig. 7.4). This may be a reflection of low maintenance energy expenditure associated with a poikilothermic stage (Kirkwood and Underwood 1984). Dronzek et al. (1986) considered that infants did not attain mature thermoregulatory capacity until 3 to 4 weeks old. An estimate of the energy requirement of cotton-top tamarins in the later stages of growth (Kirkwood and Underwood 1984) is given in Fig. 7.5.

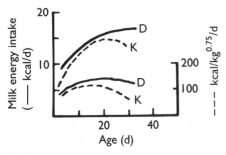

Fig. 7.4. Approximate daily energy intake provided by milk in relation to age in hand-reared infant cotton-top tamarins. The upper lines show total energy intake and the lower two show intake per metabolic weight. From Dronzek et al. (1986) (D), and Kaumanns et al. (1986) (E).

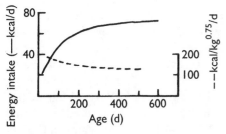

Fig. 7.5. Approximate daily energy requirement of cotton-top tamarins during growth in relation to age. The broken line shows energy requirement in relation to metabolic weight. From Kirkwood and Underwood (1984).

It is common practice to offer half-strength milk solution for the first day or two (for example, Dronzek et al. 1986). This may be beneficial if the baby has previously suckled from its mother, by making a gradual transition to the new diet. However, the possible deleterious effects of reducing energy intake should be borne in mind. There is a reasonable case for providing water, an electrolyte solution, or very dilute milk for the first one or two feeds, as this reduces the risk of aspiration pneumonia in the critical stage before the baby has learned to drink from an artificial teat.

Lactation and weaning

Cotton-tops begin to take solid food at about 30 days of age and are fully weaned at 50 to 100 days of age (Hershkovitz 1977; Cleveland and Snowdon 1984). Records in the AAZPA *Infant diet/care notebook* (Taylor and Bietz 1985) are consistent with this, as are the reports by Tardif et al. (1985) and Dronzek et al. (1986).

Dronzek et al. (1986) began weaning by adding yoghurt to Primilac at a ratio of 1:5, on days 31–34 and encouraging the infants to lap from a bowl. Gradually, other solid foods were added to the milk/yoghurt mixture: high protein baby cereal, days 34–37; egg and fruit mash, days 37–53. From day 54 the milk formula was reduced at a rate of 1 ml/d and Zu Preem Marmoset Diet was gradually introduced. Supplementary feeds of yoghurt, high protein baby cereal, mashed egg and fruit, and a vitamin supplement were offered twice daily, after the infant had been reintroduced to a family group, until 3 months of age.

Feeding

Babies have been fed by very gently expressing milk from a 1–3 ml syringe (1 ml gives greater control) fitted with either a silicone rubber rodent nipple (Hershkovitz 1977), a lacrimal catheter, or using specially constructed nipples (Ogden 1979; Dronzek et al. 1986). The latter were made by attaching a 25 gauge needle (cut to 5 mm) to a 3 ml syringe, then dipping the needle and the end of the syringe into glacial acetic acid, then latex, then glacial acetic acid again, and repeating the dipping two or three times. The syringes are then stood upright to dry for an hour, submerged in water for 24 hours, then dried again and small holes made in the tips (Ogden 1979; Dronzek et al. 1986).

Milk should be warmed to 37 °C before feeding and expressed gently using the syringe so that the baby can lap or suck at its own rate, whilst in place on a surrogate mother (Dronzek et al. 1986) or held in the hand.

For the first 8 days Dronzek et al. (1986) fed their animals at 2 hour intervals, 9 times a day (presumably with a break between about 22.00 h and 06.00 h). The feeding frequency was reduced to 6 times a day at 2.5 h intervals by day 17, 4 times a day at 4 h intervals by day 31, and to twice at 8 h intervals by day 47. Similar regimes have been reported by others.

After each feeding Dronzek et al. (1986) turned the babies on their automatic rocker (see Accommodation) for a short period to simulate carriage by a natural parent, and considered that this helped to prevent 'bloat'.

Babies can be trained to lap from a bowl by 33 days of age (Dronzek et al. 1986).

Accommodation

Neonatal cotton-tops are unable to maintain body temperature at birth. Babies have to be kept in incubators. Pook (1976) and Ogden (1979) found it necessary to keep the babies individually, because if kept together they would cling to one another and bite each others fingers and toes. Dronzek et al. (1986) placed their infants on surrogate mothers suspended in infant incubators. The temperature was maintained at 32 °C for the first two days and then lowered gradually to 27 °C during the first week. Ogden (1979) also used incubators set at 32 °C (and 70–80 per cent humidity) but maintained the temperature at this level for 4 weeks before gradually reducing it to room temperature (27 °C) by the time the babies were transferred to cages at 10–12 weeks of age. He used simple surrogates made from 5 cm stockinette packed with cotton wool, suspended at a 60 degree angle from the top of the incubator by a rubber band. The surrogates used by Dronzek et al. (1986) were made from 23 × 13 cm of wire mesh rolled into a cylinder with 50 cm of heating cable coiled inside it, and artificial brown fur wrapped over the outside. This was covered with a non-toxically tanned tamarin skin when available. The surrogate was suspended horizontally on two 30 cm elastic bands from the top of the incubator. An automated rocking system consisting of an electric motor with a reciprocating arm with a 3 cm throw (outside the incubator) to which one of the elastic suspensory bands was linked. For feeding, the infants were encouraged to climb on to another fur-covered surrogate through which, via a hole in the front, a feeding nipple could be inserted. In this colony babies remained in incubators until reaching a weight of about 150 g at 10–12 weeks of age.

Infant management notes

After each feeding, it is a well-established practice to stimulate urination and defaecation by gentle massage of the perineum with a damp cloth. Dronzek et al. (1986) found that infants began to pass urine and faeces spontaneously at about 20 days of age.

Physical development

The eyes are open at birth, and babies cling to the backs of their parents or older siblings. They usually ride over the shoulders and it is a bad sign if they are seen clinging on over the lower back or hips. This occurs if the babies are too weak to orientate normally and often precedes their falling.

Behavioural development

Babies start leaving their parents' and older siblings' backs at 22–50 days of age (Pook 1976; Cleveland and Snowdon 1984). Hand-reared babies will spontaneously leave their surrogates and begin to explore their rearing boxes at 13–17 days of age (Pook 1976; Kaumanns et al. 1986). They can use their hands to eat solid food at 34 days and groom themselves at 41 days (Pook 1976). *Saguinus oedipus* tend to be carried by their parents for longer than *Callithrix, Cebuella*, and *Leontopithecus* (Tardif et al. 1987). Tardif et al. (1985) noted that behavioural peculiarities of hand-reared infants (poor locomotion and riding each other) disappeared within 3 months of introduction to an adult pair.

Disease and mortality

Infant mortality rates have often been found to be high in captive colonies (Wolfe et al. 1975). For example, Kirkwood et al. (1985) reported that 72 (64 per cent) of 113 full-term babies (including still births) born to second generation captive-bred mothers died prior to 6 months of age. Overall percentage mortality of liveborn babies before 6 months of age reported at other colonies has also been quite high; for example, 58 (55 per cent) of 106 (Kilborn et al. 1983), 20 (55 per cent) of 36 (Evans 1983), 55 (57 per cent) of 96 (Snowdon et al. 1985), 25 (35 per cent) of 72 (Kaumanns *et al.* 1986), and 68 (40 per cent) of 172 (to 3 months, Tardif *et al.* 1984). From a survey of reports in the *International zoo yearbook*, 1967–82 Glaston *et al.* (1984) concluded the mortality during the first year of zoo-bred animals was about 45 per cent. Quite high rates of abortion and still birth have also been reported in this species in captivity, ranging from 5 per cent (Snowdon *et al.* 1985) to 20 per cent (Tardif *et al.* 1984), and 32 per cent (Kilborn *et al.* 1983).

Several factors have been shown to be associated with infant mortality, much of which occurs within the first day or week (for example, 84 per cent, Tardif *et al.* 1984) and is related to inadequate parental care or parental hostility.

Although, as indicated above, triplet litters can be common (upto 25 per cent) it is very rare for three babies to be reared. The reason for this is not known but presumably it is because there are constraints to milk production or competition for the two teats. A proportion of infant mortality can be attributed to this inability to rear triplet births.

Parity had a clear effect on infant survival at one colony, where the percentage of babies that were successfully reared increased from 18 in the first pregnancy to 71 in the fourth and fifth (Kirkwood *et al.* 1985). Snowdon *et al.* (1985) found a similar effect of parity and also that infant mortality was greater in babies born to mothers that had not had experience with care of young siblings compared with those born to mothers that had. It is possible that maternal age is also a factor and that the low infant survival of first litters is due to behavioural immaturity of young mothers. In captivity, first litters are often born at about 2.5 years of age but in the wild females may well not breed so early. Previous experience of infant care by the male probably also increases the changes of infant survival.

It has also been suggested that cage size and complexity may influence infant survival (Snowdon *et al.* 1985). One advantage of larger cages is that they can accommodate larger family groups, and older siblings can therefore be left in the family group for longer. They will then be both older and have had more experience with young siblings than those that have to be removed at 1–2 years old because of cage size constraints.

Cage design may also affect infant mortality in other ways. Babies dropped at birth will have less chance of survival if they fall onto a hard, cold surface than if the floor below is padded. If the floor is of mesh the dropped babies will cling to it and this can interfere with attempts by the parents to pick them up.

Standards of management are improving and in her survey of the large number of records of cotton-tops kept in zoos and research, Tardif (1986) found that infant survival has increased continuously from 25 per cent in 1974 to 57.8 per cent in 1983. Improvements in infant survival with time have also been recorded by Kaumanns *et al.* (1986).

Kilborn *et al.* (1983) analysed combined causes of death in infants that died during the first week of life, and their findings were: low birth weights 55 per cent, trauma 43 per cent, intracranial haemorrhage 7 per cent, pulmonary atelectasis 25 per cent, and infection 25 per cent. In animals older than 1 week but less than 1 year old, these authors found that infection was common (88 per cent) (septicaemia 62 per cent, meningitis 25 per cent, pneumonia 29 per cent) and colitis, which is very common in this species in captivity (Chalifoux *et al.* 1985) was found in 33 per cent.

Although he did not provide data for individual species, Ogden (1979) reported successfully handrearing 235 (68 per cent) out of 344 *Saguinus* and *Callithrix* babies to 30 days of age. Including only those babies that were taken soon after birth and that were apparently normal the success rate was 197 out of 228 (86 per cent) surviving to 30 days of age. Mortality was much lower after 30 days.

Preventative medicine

Infant mortality is clearly influenced by the behavioural status of the parents or family (see Diseases and mortality above), and the whole management of breeding groups must be considered when investigating causes of infant mortality (see also Accommodation).

Scullion and Terlecki (1987) reported a significantly lower incidence of neonatal mortality in infants born to captive-bred mothers when 0.25 mg diazepam in banana was given daily to each of the adult members of the family for the first week post-partum.

When triplets are born, it may be possible to rear all three by removing one baby in turn each night and providing supplementary feeding, whilst the other two remain with the parents, as has been reported in common marmosets by Ziegler et al. (1981).

This species is known to be susceptible to measles (Albrecht et al. 1980) and to *Herpesvirus hominis*, so steps should be taken to prevent these being introduced. Measles vaccination should be considered if the risk is judged to be high, but is not routinely administered in most colonies.

Often babies found abandoned after being born during the night are cold, wounded, and may be dehydrated. Initially the body temperature should be raised, then subcutaneous fluids should be given if judged to be necessary. Dronzek et al. (1986) administered a long-acting penicillin preparation to all infants with lacerations or bite wounds in addition to topical treatment of these lesions.

Ogden (1979) described the routine inclusion of ampicillin in the milk fed to hand-reared *Saguinus* and *Callithrix* species up to 7 days of age. Ideally, antibiotics should not be used unless there is a specific indication for them, and their selection should be based on identification and antibiotic sensitivity of pathogens (see Chapter 6, Preventative medicine).

Indications for hand-rearing

Babies that have been abandoned or attacked by their parents will not survive unless they are promptly removed for cross-fostering (see Reintegration) or hand-rearing. If neither of these options are possible the babies should be humanely destroyed. If no foster parents are available and if the facilities and man-hours are available, hand-rearing can be a successful way of saving babies for reintroduction into family groups (Dronzek et al. 1986). In addition to Ogden's (1979) successes reported above, 6 out of 8 babies were successfully hand-reared by Dronzek et al. (1986); 6 out of 10 by Tardif et al. (1985): and 45 out of 96 by Johnston et al. (1986), whose success rate increased from 27 to 72 per cent with experience gained over a five year period. Although we do not know of any reports of hand-reared cotton-tops successfully rearing young there seems no reason to expect any particular difficulties.

In view of the high risk of mortality in triplet litters, they should be watched closely, and as soon as one infant shows signs of weakness, for example, clinging across the lower back rather than over the shoulders, it should be removed. There is a strong case for removing one of the triplets as soon as possible after birth.

Reintegration

Cross-fostering abandoned babies to other pairs that have recently lost their own young has often been successful (Collier et al. 1981; Dronzek et al. 1986; Scullion personal communication). Collier et al. (1981) placed a baby that had been hand-reared for five days on a white towel on the floor of the cage of a pair which had stillborn twins the previous day. The colony was quietened by turning off the lights, and the female immediately reached down and picked up the baby. For 30 minutes the baby vocalized whilst the foster-parents actively transferred it between them (and dropped it several times) but then it was carried mainly on the female's back and became silent. It was seen at the teat within 8 hours and was successfully adopted.

If no lactating female is available to foster an abandoned baby, then reintroduction has to be undertaken at a later stage when the baby has been weaned and can feed itself. Dronzek et al. (1986) successfully reintroduced five out of six weanlings into family groups, and documented their technique in detail. They began by placing the 2- to 3-week-old infant in its incubator in the cage of the foster family for 2 to 3 hours each day. At 3 to 4 weeks it was placed on a surrogate in a reintroduction cage, furnished with branches for climbing, in front of the family's cage so that they could clearly see each other. Gradually, the time spent in the reintroduction cage was increased from one, to nine hours daily, but the infant was returned to the incubator at night. Later (when 4 weeks old), the reintroduction cage was placed inside the family's cage and the baby lived there day and night. When 5 to 6 weeks old, the baby was put, on its surrogate, on the roof of the family's nest box for an increasing number of the daylight hours. Close observations were made at this time for signs of aggression. The surrogate was then removed, at first for 10 minutes but for longer each day, until available only at night. At 8 to 10 weeks the infant rejected the surrogate and joined the family in the next box at night. The authors noted that reintroductions seemed to be most successful when the family groups included subadults and at least one juvenile. Tardif et al. (1985) reported gradually introducing two sets of hand-reared babies to a pair of adults with no offspring. They observed no aggression and the adults began carrying the babies 3 and 11 days after the two introductions began.

In view of the importance to future parental adequacy of subadults gaining experience with infants, cross-fostering subadults from families that have no babies to those that do or will have babies, should be considered. Collier et al. (1981) successfully transferred five out of nine to other families, but if introductions are carried out very gradually, perhaps higher success rates could be achieved.

References

Albrecht, P., Lorenz, D., Klutch, M.K., Vickers, J.H., and Ennis, F.A. (1980). Fatal measles virus infection in marmosets: pathogenesis and prophylaxis. *Infection and Immunology*, **27**, 969–78.

Brand, H.M. (1980). Influence of season on birth distribution on marmosets and tamarins. *Laboratory Animals*, **14**, 301–2.

Brand, H.M. (1981). Husbandry and breeding of a newly established colony of cotton-topped tamarins (*Saguinus oedipus*). *Laboratory Animals*, **15**, 7–11.

Chalifoux, L.V. Brieland, J.K., and King, N.W. (1985). Evolution and natural history of colonic disease in cotton-top tamarins (*Saguinus oedipus*). *Digestive Diseases and Sciences*, **30** (suppl.), 54–85.

Cicmanec, J.L., Hernandez, D.M., Jenkins, S.R., Campbell, A.K., and Smith, J.A. (1979). Hand-rearing infant Callitrichids (*Saguinus* spp. and *Callithrix jacchus*), owl monkeys (*Aotus trivirgatus*), and capuchins (*Cebus albifrons*). In *Nursery care of nonhuman primates* (ed. G.C. Ruppenthal), pp. 307–12. Plenum Press, New York.

Clapp, N.K. and Tardif, S.D. (1985). Marmoset husbandry and nutrition. *Digestive Diseases and Sciences*, **30** (suppl.), 17S–23S.

Cleveland, J. and Snowdon, C.T. (1984). Social development during the first twenty weeks in the cotton-top tamarin (*Saguinus oedipus*). *Animal Behaviour*, **32**, 432–44.

Colley, R. (1986). *Regional studbook: cotton-top tamarins*. No. 1. Penscynor Wildlife Park, West Glamorgan.

Collier, C., Kaida, S.A., and Brody, J. (1981). Fostering techniques with cotton-top tamarins *Saguinus oedipus oedipus* at Los Angeles Zoo. *International Zoo Yearbook*, **21**, 224–5.

Dronzek, L.A., Savage, A., Snowdon, C.T., Whaling, C.S., and Ziegler, T.E. (1986). Technique for hand-rearing and re-introducing rejected cotton-top tamarin infants. *Laboratory Animal Science*, **36**, 243–7.

Escajadillo, A., Bronson, R.T., Sehgal, P.K., and Hayes, K.C. (1981). Nutritional evaluation in cotton-top tamarins (*Saguinus oedipus*). *Laboratory Animal Science*, **3**, 161–5.

Evans, S. (1983). Breeding of the cotton-top tamarin *Saguinus oedipus oedipus*: a comparison with the common marmoset. *Zoo Biology*, **2**, 47–54.

Foster, B. (1985). In *Infant diet/care notebook* (ed. S.H. Taylor and A.D. Bietz) American Association of Zoo Parks and Aquariums, Wheeling, Virginia.

Gengozian, N., Batson, J.S., and Smith, T.A. (1978). Breeding marmosets in a colony environment. *Primates in Medicine*, **10**, 71–8.

Glatston, A.R., Geilvoet-Soeteman, E., Hora-Pecek, G., and van Hoof, J.A.R.A.M. (1984). The influence of the zoo environment on social behaviour of groups of cotton-top tamarins *Saguinus oedipus oedipus*. *Zoo Biology*, **3**, 241–53.

Hershkovitz, P. (1977). *Living new world monkeys (Platyrrhini)*, Vol. 1. University of Chicago Press.

IUCN, (1990). *1990 IUCN red list of threatened animals*. pp. 11. IUCN, Gland, Switzerland.

Jenness, R. and Sloan, R.E. (1970). The composition of milks of various species: a review. *Dairy Science Abstracts*, **32**, 599–612.

Johnson, L.D., Petto, A.J., Boy, D.S., Sehgal, P.K., and Beland, M.E. (1986). The effect of perinatal mortality on colony-born production at the New England Regional Primate Research Center. In (ed. K. Benirschke), *Primates. The road to self-sustaining populations* pp. 771–9. Springer-Verlag, New York.

Kahn, D. (1985). In *Infant diet/care notebook* (ed. S.H. Taylor and A.D. Bietz). American Association of Zoo Parks and Aquariums, Wheeling, Virginia.

Kaumanns, W., Klensang, H., Oftenbuttell, T., Rohrhuber, B., and Scwibbe, M. (1986). Zur haltung von Lisztaffen (*Saguinus oedipus oedipus*). *Zeitschrifte des Kölner Zoo*, **2**, 43–62.

Kilborn, J.A., Sehgal, P.K., Johnson, L.D., Beland, M., and Bronson, R.T. (1983). A retrospective study of infant mortality of cotton top tamarins (*Saguinus oedipus*) in captive breeding. *Laboratory Animal Science*, **33**, 168–71.

Kirkwood, J.K. (1983). Effects of diet on health, weight and litter size in captive cotton-top tamarins *Saguinus oedipus oedipus*. *Primates*, **24**, 515–20.

Kirkwood, J.K. and Underwood, S.J. (1984). Energy requirements of captive cotton-top tamarins (*Saguinus oedipus oedipus*). *Folia Primatologia*, **42**, 180–7.

Kirkwood, J.K., Epstein, M.A., and Terlecki, A.J. (1983). Factors influencing population growth of a colony of cotton-top tamarins. *Laboratory Animals*, **17**, 35–41.

Kirkwood, J.K., Epstein, M.A. Terlecki, A.J., and Underwood, S. (1985). Rearing a second generation of cotton-top tamarins *Saguinus oedipus oedipus* in captivity. *Laboratory Animals*, **19**, 269–72.

Neyman, P.F. (1977). Aspects of the ecology and social organisation of free-ranging cotton-top tamarins (*Saguinus oedipus*) and the conservation status of the species. In *The biology and conservation of the Callitrichidae* (ed. D.G. Kleiman), pp. 39–71. Smithsonian Institution Press, Washington D.C.

Ogden, J.D. (1979). Hand-rearing *Saguinus and Callithrix genera* of marmosets. In *Nursery care of nonhuman primates* (ed. G.C. Ruppenthal), pp. 313–9. Plenum Press, New York.

Pook, A.G. (1976). Some notes on the development of hand-reared infants of four species of marmoset (Callitrichidae). *Jersey Wildlife preservation Trust Annual Report*, **13**, 38–46.

Price, E.C. (1988). Birth and perinatal behaviour in captive cotton-top tamarins (*Saguinus o. oedipus*). *International Journal of Primatology*, **8**, 482.

Schmidt, E. and Solder, J. (1985). In *Infant diet/care notebook* (ed. S.H. Taylor and A.D. Bietz) American Association of Zoo Parks and Aquariums, Wheeling, Virginia.

Scullion, F.T. and Terlecki, A.J. (1987). The use of diazepam for increasing infant survivability in the cotton top tamarin (*Saguinus o. oedipus*). *Proceedings of the First International Conference on Zoological and Avian Medicine*, pp. 501–11. American Association of Zoo Veterinarians.

Snowdon, C.T., Savage, A., and McConnell, P.B. (1985). A breeding colony of cotton-top tamarins (*Saguinus oedipus*). *Laboratory Animal Science*, **35**, 477–80.

Tardif, S.D. (1986). Status of the endangered cotton-top tamarin (*Saguinus oedipus*) in captivity. *Primate Report*, **14**, 239–40.

Tardif, S.D., Richter, C.B., and Carson, R.L. (1984). Reproductive performance of three species of Callitrichidae. *Laboratory Animal Science*, **34**, 272–5.

Tardif, S.D., Lenhard, A., Carson, R., and McArthur, A. (1985). Hand-rearing of infant *Saguinus oedipus* with subsequent introduction into social groups. *American Journal of Primatology* **8**, 368–9.

Tardif, S.D., Harrison, M.L., and Simek, M.A. (1987). Infant care in marmosets and tamarins. *International Journal of Primatology*, **8**, 436.

Taylor, S.H. and Bietz, A.D. (ed.) (1985). *Infant diet/care notebook*. American Association of Zoo Parks and Aquariums, Wheeling, Virginia.

Wolfe, L.G., Deinhardt, F., Ogden, J.D., Adams, M.R., and Fisher, L.E. (1975). Reproduction of wild-caught and laboratory-born marmoset species used in biomedical research (*Saguinus* spp: *Callithrix jacchus*). *Laboratory Animal Science*, **25**, 802–13.

Ziegler, T.E., Stein, F.J. Sis, R.F., Coleman, M.S., and Green, J.H. (1981). Supplemental feeding of marmoset (*Callithrix jacchus*) triplets. *Laboratory Animal Science*, **31**, 194–5.

Golden lion tamarin

8 Golden lion tamarin

Species
The golden lion tamarin
Leontopithecus rosalia

ISIS No. 1406007003001001

Status, subspecies, and distribution
There are three subspecies of *Leontopithecus r. rosalia*, the golden lion tamarin, *L.r. chrysomelas*, the golden-headed lion tamarin, and *L.r. chrysopygus*, the golden-rumped or black lion tamarin. These species are restricted to the Atlantic forest of Brazil, and their ranges have been reduced to tiny isolated fragments (Mittermeier 1986). All three subspecies are endangered (Mittermeier *et al.* 1986, IUCN 1990). *Leontopithecus r. chrysopygus* is the rarest and Mittermeier (1986) considered that the wild population was unlikely to exceed 100. Coimbra-Filho and Mittermeier (1977) estimated the number of wild populations of *L.r. chrysomela* to be 100 to 200, and Kleiman *et al.* (1986) suggested that there were 250 to 350 wild *L.r. rosalia*.

In the early 1970s, there were thought to be less than 80 *L.r. rosalia* remaining in captivity (Kleiman *et al.* 1982), but the species has bred well in captivity. This population reached 370 by the end of 1983, and thereafter grew at 20 to 25 per cent a year (Kleiman *et al.* 1986). The breeding of the captive population has been controlled to balance the genetic contribution from the founder stock.

The species is not used for biomedical research.

Sex ratio
Kleiman *et al.* (1982) reported that at birth there were significantly more males than females (273:210 in their sample), but that infant mortality was higher among males.

Social structure
Groups seen in the wild range from 4 to 11 individuals, and it appears that there is only one breeding female in each group (Kleiman *et al.* 1986). Mothers are the principal carriers of the infants for the first three weeks but thereafter fathers take the main share of this duty. Older siblings also assist in infant carrying (Hoage 1982).

Breeding age
In captivity, the mean age at first conception was 29.3 months, and the age for first insemination (males) was 28.3 months (Kleiman *et al.* 1986). These figures are likely to be influenced by the age at pairing, and sexual maturity can occur earlier (Kingston 1969; Eisenberg 1977). Conception has been recorded at 14 months (Kleiman and Jones 1977).

Longevity
The 1982 *International studbook* for the species records the death of both a male and a female at 15 years of age (Kleiman and Evans 1983).

Seasonality
In captivity, births have been recorded in every month of the year and between about 15 and 45 per cent of females produce two litters a year (Kleiman *et al.* 1982). Amongst those in captivity in the northern hemisphere, a seasonal pattern is seen, with a tendency for litters to be born between March and August. In Brazil, the main breeding season is late in the year (Coimbra-Filho and Mittermeier 1973). Because of the relatively short gestation period, compared with other Callitrichidae, it is possible for captive females to produce three litters within one year, and a proportion do so (Kleiman *et al.* 1982).

Gestation

The mean gestation period has been determined at 129 days (Wilson 1977).

Pregnancy diagnosis

No specific data are available. See the relevant sections for other Callitrichidae.

Birth

Birth usually occurs at night (Kingston 1969).

Litter size

The litter size varies from 1 to 4, with twins being the most common (typically 60 to 80 per cent). As observed in other Callitrichidae maintained in captivity, there has been an increase in the number of triplet litters (up to about 20 per cent). Unlike *Callithrix* and *Saguinus* species, *Leontopithecus* is often able to rear all three (Kleiman *et al.* 1982).

Adult weight

Adult weight varies with the system of management for reasons that are not clear. Kleiman *et al.* (1982) reported that records of wild-caught and captive-bred animals prior to 1970 were typically in the 500 to 650 g range, but more recent records of captive animals were in the 650 to 750 g range. Hoage (1982) reported mean adult weights of 715 g and 703 g for males and females respectively.

Neonate weight

The mean weight of 11 babies was found by Hoage (1982) to be 60.6 g, with a range from 52.1 to 74.6 g. Neonate weight is related to litter size, with singletons weighing about 70 g, twins about 60 g, and triplets about 50 g each (Hoage 1982; Wilson 1977).

Adult diet and energy requirements

Diets fed to golden lion tamarins in captivity are generally similar to those for cotton-top tamarins and common marmosets. The daily metabolizable energy requirement of an adult weighing 700 g is likely to be about 100 kcal per day.

Growth

Hoage (1982) provided growth curves for a small sample of males and females. His animals grew at about 2.2 g/d from 8 to 32 weeks of age and thereafter growth averaged about 0.5 g/d. The animals reached about 600 g at 1 year of age (Fig. 8.1). The growth of females may lag slightly behind that of males, but more data are needed to confirm this. Adult weight is reached at between 15 and 20 months (Hoage 1982). One male whose growth was recorded by Hoage (1982) reached 600 g at about 8 months.

The weight gains of two babies hand-reared by Dumond *et al.* (1979) are shown in Fig. 8.2. Although there are many reports of hand-rearing in the *International studbook* (Kleiman and Evans 1983), there is little published information on techniques used and performance of the babies.

Milk and milk intake

A sample of milk 'collected from a female 3 days post-partum was found to contain 5.8 per cent fat, 5.7 per cent protein, 6.9 per cent lactose, and 0.78 per cent ash (Buss 1975). On a dry matter basis this corresponds to 30 per cent, 30 per cent, 36 per cent, and 4 per cent, of fat, protein, lactose, and ash respectively. The metabolizable energy value of the fresh milk, deduced from this composition, is about 0.71 kcal/ml. Buss (1975) also measured the concentration of some minerals and the relative proportions of fatty acids.

The protein content of the milk, as measured by Buss (1975), is considerably higher than that of human milk replacers, but the latter have been used successfully for hand-rearing. For example, Dumond *et al.* (1979), who indicated that they were following the example of Oklahoma City Zoo where golden lion tamarins had been successfully reared (Rohrer 1979), used Similac with Iron.

The milk formula listed by the Golden Lion Tamarin Committee (1990) was SMA (12.8 g per 100 ml water) for the first 2 days, to which an increasing amount of Esbilac portion is added (up to 3.5 g per 100 ml formula) by day 5.

On the first day Dumond *et al.* (1979) gave a total of 5 ml, but this increased to 18–20 ml on the second day and milk intake reached a plateau of about 25–30 ml by the fifth day. Thereafter, milk intake remained at around this level until the babies were 21 days old then began to drop

Golden lion tamarin

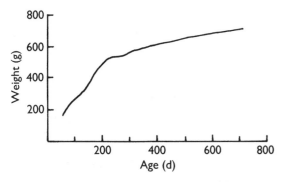

Fig. 8.1. Average growth curve of the golden lion tamarin. From Hoage (1982).

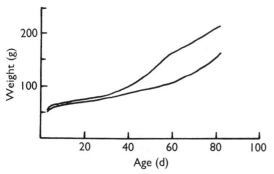

Fig. 8.2. Neonatal growth rates of two hand-reared golden lion tamarins. From Dumond *et al.* (1979).

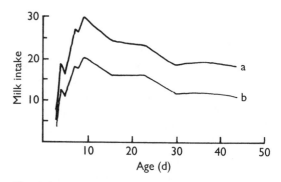

Fig. 8.3. Approximate daily (a) milk, ml/d, and (b) energy intake, kcal/d, provided by the milk of hand-reared golden lion tamarins. From Dumond *et al.* (1979).

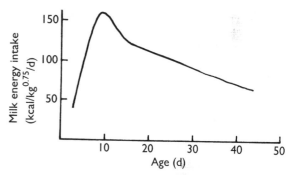

Fig. 8.4. Approximate daily energy intake provided by milk in relation to metabolic weight in golden lion tamarins. From Dumond *et al.* (1979).

gradually as the intake of solid food increased (Fig. 8.3). The metabolizable energy provided by the milk rose from 13 kcal on the day 2 to a peak of about 20 kcal/d on day 7 (Fig. 8.3). From days 2 to 21, the energy intake was in the range 90–160 kcal/d per unit metabolic weight (Fig. 8.4). A baby reared by Rohrer (1985) had an average daily Similac intake of 24 ml, 32 ml, and 48 ml, at 1 week, 1 month, and 2 months of age respectively.

Lactation and weaning

Mother-reared golden lion tamarins start taking some solid food at 4 to 5 weeks of age, while being carried. They dismount to feed alone by 8 weeks of age and weaning is completed at 11 to 15 weeks of age (Hoage 1982). If the milk energy intake observed in the babies hand-fed by Dumond *et al.* (1979) is typical, then a mother rearing triplets may be producing 45 kcal of milk per day at peak lactation. If the efficiency of milk synthesis is about 70 per cent (MacDonald *et al.* 1981), the mother will have to eat about 65 kcal per day above her own maintenance requirement; that is, a total of about 1.65 times her maintenance requirement.

Feeding

Dumond *et al.* (1979) fed their baby golden lion tamarins with a syringe. The babies were fed nine, decreasing to eight times a day during the first week, and decreasing to four or five times a day by 6 weeks of age. At 56 days of age the babies were no longer hand-fed but continued to take milk from a pan. Rohrer (1979) fed her baby golden lion tamarins every 2 hours, round the clock, for the first 3 to 4 weeks. She fed them by expressing milk from a 1 ml syringe for them to lap, rather than using teats.

In some cases, Rohrer (1979) added small amounts of Gerbers baby cereal to the Similac when the babies were 3 weeks old, and also began to offer strained fruits by syringe at this time. However, Rohrer recommended continuing to syringe-feed with Similac whilst making fruit, crickets, and marmoset-pellets available to the infants.

There appears to be little specific literature on the feeding of infants of this species but the information presented elsewhere for common marmosets and cotton-top tamarins is relevant.

Accommodation

Rohrer (1979) placed babies in a human infant incubator, lined with towels, and the air temperature regulated at 29–32 °C. Babies were given pieces of towelling to cling to. The practices described for other Callitrichidae are relevant to this species.

Infant management notes

The twins hand-fed by Dumond et al. (1979) were left with their parents between feeds. They were hand-fed because two previous litters born to the mother had died through apparent starvation, and it was thought that the mother either failed to produce milk or did not allow them to feed. Expecting that this would happen again, when the female next became pregnant efforts were made to habituate her to regular handling. After birth no great difficulties were experienced removing the babies for hand-feeding or returning them after each feed.

Techniques described for artificial rearing of other Callitrichidae are applicable to this species. The Golden Lion Tamarin Management Committee (1990) stated that infants should not be hand-reared without their studbook keeper's permission, and that hand-rearing is usually recommended only for genetically valuable animals.

Physical development

There is scant literature on this subject for the golden lion tamarin. Hoage (1982) provides some data on tail length and head to rump length during growth. His data show that total length increases from about 24 cm at birth, to 40 cm at 3 months, and to an adult size of about 55 cm at 12 to 18 months.

Behavioural development

Infants have been first observed off and independent of their carriers at about 25 days (Altmann-Schonberner 1965), and begin tasting solid foods at about this time. They cease to be regularly carried at about 11 weeks of age (Hoage 1982). Grooming attempts are seen at about 3 weeks of age. The mean age when scent-marking is first observed is about 17 weeks. Juveniles of 35 weeks carry younger siblings (Hoage 1982).

Further details of behavioural development are given by Hershkovitz (1977) and Hoage (1982).

Disease and mortality

Kleiman et al. (1982) reported that analysis of the studbook records at that time indicated that infant and juvenile mortality (all deaths prior to 1 year of age) averaged 40–45 per cent, but had been 50–60 per cent between 1970 and 1975. Abortions are not always observed or recorded but their data indicated that a minimum of 12 per cent of pregnancies did not reach term.

As far as we are aware there have been no detailed investigations into the causes of abortion or neonatal mortality in this species, but poor mothering, resulting from removal from the family group too early for normal behavioural development and experience in caring for young siblings, is likely to be one factor (Kleiman 1977).

Preventative medicine

As in other species of Callitrichidae maintained in captivity, high rates of infant mortality are probably often related to management practices. Juveniles should be left in their family groups long enough to gain experience in carrying siblings before being paired for breeding.

Golden lion tamarins may be susceptible to fatal infection with *Herpesviru hominis*, so they should be isolated from anyone with herpes 'cold sores'. Preventative measures listed for other Callitrichidae apply to this species.

Indications for hand-rearing

There is no specific information.

Reintegration

Hand-reared golden lion tamarins were apparently successfully reintegrated into family groups by Rohrer (1979), who placed hand-reared infants in a cage within the cage of the group, for increasing periods daily. After 4 to 6 weeks, the infant was let out to join the other members of the group. Rohrer did not report any great difficulties with reintegration of hand-reared babies, but no data are available on subsequent breeding success.

The Golden Lion Tamarin Management Committee (1990) strongly recommended reintroduction of hand-reared infants to family groups at no later than 3-4 weeks to ensure adequate socialisation, and that prior to this socialization can be provided by placing the incubator or rearing cage inside the family's cage during the day.

References

Altmann-Schonberner, D. (1965). Beobachtungen über Aufzucht und Entwicklung des Verhaltens beim grossen Löwenäffchen, *Leontocebus rosalia*. *Der Zoologische Garten*, **31**, 227-39.

Buss, D.H. (1975). Composition of milk from a golden lion marmoset. *Laboratory Primate Newsletter*, **14**, 17-18.

Coimbra-Filho, A.F. and Mittermeier, R.A. (1973). Distribution and ecology of the genus Leontopithecus Lesson 1840 in Brazil. *Primates*, **14**, 47-66.

Coimbra-Filho, A.F. and Mittermeier, R.A. (1977). Conservation of the Brazilian lion tamarins (*Leontopithecus rosalia*). In *Primate conservation* (ed. Prince Rainier of Morocco and G.H. Bourne), pp. 59-94. Academic Press, New York.

Dumond, F.V., Hoover, B.L., and Norconk, M.A. (1979). Hand-feeding parent-reared Golden lion tamarins *Leontopithecus rosalia rosalia* at Monkey Jungle. *International Zoo Yearbook*, **19**, 155-8.

Eisenberg, J.F. (1977). Comparative ecology and reproduction of New World monkeys. In *Biology and conservation of the callitrichidae* (ed. D.G. Kleiman), pp. 13-22. Smithsonian Institution Press, Washington D.C.

Golden Lion Tamarin Management Committee (1990). Husbandry protocol for Golden Lion Tamarins (*Leontopithecus rosalia rosalia*). National Zoo, Washington D.C.

Hershkovitz, P. (1977). *Living New World monkeys (Platyrrhini)*, Vol. 1, pp. 847-63. University of Chicago Press.

Hoage, R.J. (1982). Social and physical maturation in captive lion tamarins *Leontopithecus rosalia rosalia* (Primates: Callitrichidae). *Smithsonian Contributions to Zoology*, **354**, 1-56.

IUCN (1990). *1990 IUCN red list of threatened animals*. pp. 11. IUCN. Gland, Switzerland.

Kingston, W.R. (1969). Marmosets and tamarins. In *Hazards of handling simians. Laboratory Animal handbooks*, Vol. 4 (ed. F.T. Perkins and P.N. O'Donoghue), 243-50. Laboratory Animals Ltd, Newbury, UK.

Kleiman, D.G. (1977). Progress and problems in lion tamarin *Leontopithecus rosalia rosalia* reproduction. *International Zoo Yearbook*, **17**, 92-7.

Kleiman, D.G. and Evans, R.F. (1983). *1982 International studbook for the golden lion tamarin* Leontopithecus rosalia rosalia. National Zoological Park, Washington D.C.

Kleiman. D.G. and Jones, M. (1977). The current status of *Leontopithecus rosalia* in captivity with comments on breeding success at the National Zoological Park. In *Biology and conservation of the callitrichidae* (ed. D.G. Kleiman), pp. 215-18. Smithsonian Institution Press, Washington D.C.

Kleiman, D.G., Ballou, J.D., and Evans, R.F. (1982). An analysis of recent reproductive trends in captive Golden lion tamarins *Leontopithecus r. rosalia* with comments on their future demographic management. *International Zoo Yearbook*, **22**, 94-101.

Kleiman, D.G., Beck, B.B., Dietz, J.M., Dietz, L.A., Ballou, J.D., and Coimbra-Filho, A.F. (1986). Conservation program for the golden lion tamarin: captive research and management, ecological studies, educational strategies, and reintroduction. In *Primates. The road to self-sustaining populations* (ed. K. Benirschke), pp. 959-79. Springer-Verlag, New York.

MacDonald, P., Edwards, R.A., and Greenhalgh, J.F.D. (1981). *Animal nutrition*. Oliver & Boyd, Edinburgh.

Mittermeier, R.A. (1986). Primate conservation priorities in the Neotropical region. In *Primates. The road to self-sustaining populations* (ed. K. Benirschke), pp. 221-40. Springer-Verlag, New York.

Mittermeier, R.A., Oates, J.F., Eudey, A.E., and Thornback, J. (1986). Primate conservation. In *Behaviour, conservation, and ecology. Comparative Primate Biology 2A* (ed. G. Mitchell, and J. Erwin), pp. 3-72. Alan R. Liss, New York.

Rohrer, M.A. (1979). Hand-rearing golden lion marmosets, *Leontopithecus rosalia*, at the Oklahoma City Zoo. *Animal Keepers Forum*, **6**, 33-39.

Rohrer, M.A. (1985). Golden lion tamarin. In *Infant diet/care notebook* (ed. S.H. Taylor and A.D. Bietz). American Association of Zoo Parks and Aquaria. Wheeling, Virginia.

Wilson, C.G. (1977). Gestation and reproduction in golden lion tamarins. In *Biology and conservation of the callitrichidae* (ed. D.G. Kleiman), pp. 191-2. Smithsonian Institution, Washington D.C.

Owl monkey (© Zoological Society of London)

9 Owl monkey (night monkey or douroucouli)

Species
The owl monkey (night monkey or douroucouli) *Aotus trivirgatus*

ISIS No.   1406006001001

Status, subspecies, and distribution
The geographical range of the owl monkey extends from northern Argentina, through the countries of central South America: Venezuela, Brazil, Bolivia, Paraguay, Peru, Equador, and Colombia to Panama (Wolfheim 1983). It is a nocturnal primate which inhabits many types of forest (Wolfheim 1983; Wright 1981; Meritt 1980).

There are some morphological differences between races from various parts of the range, but the owl monkey is considered to be one species. However, karyotyping has revealed wide variation in the composition of the chromosomes from individuals from different geographical areas (Ma *et al.* 1976). Offspring resulting from mating between individuals of different karyomorphs are likely to be infertile (de Boer 1982).

The species was considered by Wolfheim (1983) to be in little danger of extinction, although its range was decreasing in Argentina. It is not presently classified as threatened (IUCN 1990).

Owl monkeys are quite widely kept in zoos where the first recorded captive breeding was at San Diego in 1937 (Jones 1986). Flesness (1986) in a review of ISIS census data estimated the total population in ISIS-registered establishments in 1984 to be 142, and 133 were ISIS-registered as of December 1988 (ISIS 1989). They have been used in biomedical research especially for malaria and eye research. Johnsen and Whitehair (1986) estimated the number of potential and actual breeding animals in principal research establishments in the United States in 1984, to be 390, with 40 live births that year.

Sex ratio
The sex ratio at birth is about 1:1 (Meritt 1980; Asakura and Okada 1972).

Social structure
Owl monkeys live in small family groups, of two to seven individuals, consisting of a breeding pair and their young (Wolfheim 1983; Dixson 1982; Wright 1981).

Breeding age
Puberty is reached by the male at 8 to 10 months, and probably by the female at 18 to 24 months of age (Dixson 1982). The offspring do not exhibit sexual behaviour until they are removed from the family group (Dixson 1987).

Longevity
The maximum lifespan is 19 years or more (Meritt 1980).

Seasonality
No seasonal pattern of births has been observed in captivity (Dixson 1982; Meritt and 1980; Elliott *et al.* 1976), but among wild owl monkeys in Peru, there may be a peak in birth rate during the rainy season (Wright 1981). The duration of the ovarian cycle is 15–16 days (Dixson *et al.* 1980).

Gestation
Meritt (1977) estimated the gestation period at 155 days, but more recent evidence suggests that it is shorter. Hall and Hodgen (1979) considered it to average 126 days and Hunter *et al.* (1979) estimated 133 days. The interbirth interval in a

captive colony was found to vary from 166 to 419 days, with a mean of 253 days (Dixson 1982).

Pregnancy diagnosis

Abdominal distension is only noticeable in late pregnancy (Meritt 1980). During the last 100 days of pregnancy the Subhuman Primate Pregnancy Test (National Institute of Health, USA) for urinary chorionic gonadotrophin can be used (Dixson 1982). It is likely that pregnancy could also be detected by palpation or ultrasonography from a month after conception.

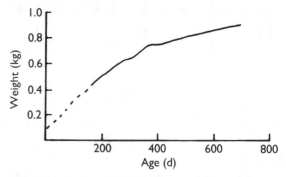

Fig. 9.1. Average growth curve of the owl monkey. From Dixson (1987).

Birth

In this nocturnal primate birth occurs during the daylight hours (Meritt 1980). In one account of birth it was recorded that birth occurs 10 to 20 minutes after the first appearance of blood at the vulva. The placenta is passed about 15 minutes after birth.

Litter size

The usual litter size is one (Dixson 1982), but twins occur occasionally (Wright 1981; Merritt 1980).

Adult weight

Individuals of both sexes weigh between 800 and 1080 g when adult (Dixson 1982) with means close to 900 g (Elliott et al. 1976).

Neonate weight

Dixson (1982) reported the normal birth weight to be 90 to 105 g. The mean weight of 36 apparently full-term babies recorded by Hall et al. (1979) was 90 g, and the range was 69 to 114 g. A mean of 90 g was also reported for colony-born owl monkeys by King and Chalifoux (1986).

Adult diet

Analysis of the stomach contents of wild owl monkeys indicated that the diet comprises 65 per cent fruit, 30 per cent foliage, and 5 per cent invertebrate and vertebrate animal prey (Hladik et al. 1971). In captivity, diets usually include a commercial primate pellet, supplemented with a variety of fruit, vegetables, and protein sources, such as locusts, mealworms, and hard-boiled egg (Dixson 1982; Meritt 1980; Asakura and Okada 1972). Supplementary vitamin and mineral preparations are also generally included. Haemolytic anaemia responsive to vitamin E therapy has been reported (Sehgal et al. 1980), so it is wise to ensure adequate levels of this vitamin.

Growth

A typical growth curve, based on data for mother-reared captive young presented by Dixson (1987) is shown in Fig. 9.1 (see also Hall et al. 1979). During the first 6 months the growth rate is roughly linear and averages about 2.6 g/d. Adult weight is reached by about 2 years of age. The average growth curves reported for hand-reared infants by Meritt and Meritt (1978) and Cicmanec et al. (1979) were very similar. Both show a linear gain from 90 g at birth to about 230 g at 70 days.

Milk and milk intake

As far as we are aware there are no data on the milk composition of this species. Human milk replacers have been used successfully for hand-rearing. Meritt and Meritt (1978) used Similac with Iron, and Cicmanec et al. (1979) used a formula comprised of 15 g Sustagen and 15 g SMA powders per 60 ml water. The latter had an energy density of 1.6 kcal/ml. Anderson (1986) recommended Similac for hand-rearing owl monkeys.

The milk intake of hand-reared owl monkeys described by Meritt and Meritt (1978) is shown in Fig. 9.2. The intake of their animals increased from 10 ml on the first day to 20 ml by 7 days, and to 50 ml by the third week. This indicates a comparatively low energy intake rate (less than 100

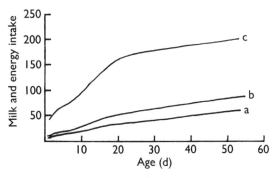

Fig. 9.2. Approximate daily milk in ml/d (a), energy intake provided by milk in kcal/d (b), and (c) energy intake in relation to metabolic weight (kcal/kg$^{0.75}$/d) of hand-reared owl monkeys. From G.F. Meritt and D.A. Meritt (1978).

kcal/d per metabolic weight) during the first week or two. The daily milk intake recommended by Cicmanec et al. (1979) was similar to that shown in Fig. 09.2, but the energy density of the milk they used was twice as high, so the energy intakes of their infants were greater. In spite of this, the rate of growth of the infants of Cicmanec et al. (1979) were very similar to that of those reared by Meritt and Meritt (1978).

Lactation and weaning

Infants first suck within 2 to 7 hours of birth (Meritt 1980). During the first month babies suck about 20 times per day and this declines to 3 or 4 times a day by the eighth month (Wright 1981). Although sucking has been observed until the ninth month after parturition (Wright 1981), weaning can be completed by 12 to 14 weeks in captivity (Meritt 1980).

Feeding

Meritt and Meritt (1978) hand-reared infant owl monkeys using a 25 ml glass bottle fitted with a long rubber nipple. They initially fed the babies at 2–2.5 hour intervals between 07.00 h and 23.00 h, but the feeding rate was reduced to 4 feeds per day by the eighth week. Solid food (peeled apple) was first offered during the third week, and a dish of diced fruit and vegetables was offered at the time of each milk feed during the fourth week. Monkey chow was crumbled over the fruit when the babies reached 6 weeks of age. Bottle-feeding was discontinued at 5 months but milk, with added baby cereal, was still made available in a dish. Meritt (1980) noted, from observation of one infant, that solid foods were first eaten on day 36 and were regularly eaten by day 49.

Accommodation

Cicmanec et al. (1979) kept their infants in incubators at 31–32 °C and 50 per cent humidity. They mentioned that the thermoregulatory ability of New World primates is not well developed at birth. Meritt and Meritt (1978) maintained incubator temperature at 27 °C and gradually reduced it after 4 weeks.

Infant management notes

Infant owl monkeys are carried by their parents and older siblings. A surrogate of the type described for hand-rearing squirrel monkeys (Chapter 10) would be suitable. Stationary surrogates have been reported to work well in this species (Anderson 1986), but there is no specific published information.

Physical development

At birth, the babies are covered with dense, short, silky fur over the head, neck, and back, but rather sparsely covered on the ventral surfaces. The eyes are open from birth.

The deciduous teeth erupt as follows: upper and lower central incisors, 2.3 weeks; lower second incisors, 2.9 weeks; upper second incisors and upper and lower first premolars, 3.5 weeks; upper and lower second premolars, 3.8 weeks; upper and lower canines, 4.6 weeks; and third premolars, 5.6 weeks (Hall et al. 1979). The permanent teeth appear between 4 and 13 months of age as follows: upper and lower first molars, 4.9 and 4.3 months respectively; upper and lower second molars, 7.3 and 6.4 months; first incisors, 9.5 months; lower third molar, 9.9 months; second incisors, 10.7 months; upper and lower third premolars, 11.5 and 11.0 months; upper third molars, 11.2 months; upper and lower second premolars, 11.9 and 12.3 months; first premolars, 12.4 months; and upper and lower canines, 15 and

14 months (Hall *et al.* 1979). Minor variations in the sequences are observed.

At puberty the subcaudal scent-marking gland on the ventral surface of the base of the tail grows (Dixson 1987). This process begins at 280–370 days of age and is complete at 340–440 days of age.

Behavioural development

During the first few days babies cling to their mothers in a characteristic position between the thigh and hip (Wright 1981; Dixson 1982). The grasping reflex is strong from birth, and babies can crawl over their mothers to find the teats. From about 1 week of age the infant is mostly carried by the father, but older siblings also assist (Dixson 1982). At about 4 weeks of age the infants begin independent exploration, and by 55 days of age infants can jump about half a metre (Meritt 1980). At 2 months of age wild owl monkey infants spend about 93 per cent of the time on their fathers, 2 per cent on their mothers, and 4 per cent by themselves, but by 3 months of age they spend 71 per cent of the time by themselves (Wright 1981).

Diseases and mortality

Meritt and Meritt (1978) found that infants which weighed less than 80 g at birth had a low survival rate, and also that primiparous females were more likely than others to reject or attack their babies. Like other New World primates, experience in carrying young siblings may have a beneficial effect on subsequent maternal or paternal behaviour. Juveniles should therefore be left within the family group until they are about 2 years old. Parental mutilation is listed as a common cause of death — 15 (19 per cent) of 80 — in colony-born owl monkeys by King and Chalifoux (1986), and is also mentioned by Cicmanec *et al.* 1979. The latter authors also reported that candidiasis and pneumonia were the most common clinical problems they encountered in infant New World primates.

Of 203 births at the New England Primate Research Center 91 (45 per cent) were stillborn. Of the 112 live births, 20 (18 per cent) died prior to 30 days of age, and a further 50 (45 per cent) died before reaching 2 years of age (Johnson *et al.* 1986). The high rate of still births in this colony was shown by Ma *et al.* (1976) to be due to breeding between individuals of different karyotypes (see also Rouse *et al.* 1981).

Owl monkeys are susceptible to infection by *Herpesvirus hominis*, the virus that causes 'cold sores' in man, and epidemics with high mortality have occurred (Melendez *et al.* 1969).

Preventative medicine

Owl monkeys should be kept isolated from people with *Herpesvirus hominis* lesions. Like other New World primates they probably require vitamin D_3. This is present in most milk replacers in adequate concentrations.

Indications for hand-rearing

Maternal illness or neglect are circumstances in which a baby should be taken for hand-rearing or euthanasia. The chances of survival of weak or premature babies can be increased by artificial rearing.

Following the death of the mother, a baby has been successfully reared by leaving it with its father but removing it regularly for feeding (Anderson 1986).

Reintegration

Meritt and Meritt (1978) reported no difficulties in pairing hand-reared young with other hand- or mother-reared individuals. There do not appear to be any data on subsequent breeding performance. However, it is likely that using techniques for early reintegration such as those described for squirrel monkeys and cotton-top tamarins, normal behavioural development will not be compromised.

References

Anderson, J.H. (1986). Rearing and intensive care of neonatal and infant nonhuman primates. In *Primates. The road to self-sustaining populations* (ed. K. Benirschke), pp. 747–62. Springer-Verlag, New York.

Asakura, S. and Okada, S. (1972). Breeding up douroucoulis *Aotus trivirgatus* at Tama Zoo, Tokyo. *International Zoo Yearbook*, **12**, 47–8.

Cicmanec, J.L., Hernandez, D.M., Jenkins, S.R., Campbell, A.K., and Smith, J.A. (1979). Hand-rearing infant callitrichids (*Saguinus* spp. an *Callithrix jacchus*), owl monkeys (*Aotus trivirgatus*), an capuchins (*Cebus albifrons*). In *Nursery care of nonhuman primates* (ed. G.C. Ruppenthal), pp. 309–11. Plenum Press, New York.

de Boer, L.E.M. (1982) Karyological problems in breeding owl monkeys *Aotus trivirgatus*. *International Zoo Yearbook*, **22**, 119–24.

Dixson, A.F. (1982). Some observations on the reproductive physiology and behaviour of the owl monkey *Aotus trivirgatus* in captivity. *International Zoo Yearbook*, **22**, 115–19.

Dixson, A.F. (1987). The owl monkey. In *The UFAW handbook on the care and management of laboratory animals* (ed. T.B. Poole), pp. 582–5. Longman, Harlow.

Dixson, A.F. Martin, R.D., Bonney, R.C., and Flemming, D. (1980). Reproductive biology of the owl monkey, *Aotus trivirgatus griseimembra*. In *Non-human primate models for the study of human reproduction* (ed. T.C.A. Kumar), pp. 61–8. Karger, Basel.

Elliot, M.W., Sehgal, P.K., and Chalifoux, L.V. (1976). Management and breeding of *Aotus trivirgatus*. *Laboratory Animal Science*, **26**, 1037–40.

Flesness, N.R. (1986). Captive status and genetic considerations. In *Primates. The road to self-sustaining populations* (ed. K. Benirschke), pp. 845–56. Springer-Verlag, New York.

Hall, R.D., and Hodgen, G.D. (1979). Pregnancy diagnosis in owl monkeys (*Aotus trivirgatus*): evaluation of the haemagglutination inhibition test for urinary chorionic gonadotrophin. *Laboratory Animal Science*, **29**, 345–8.

Hall, R.D., Beattie, R.J., and Wyckoff, G.H. Jr. (1979). Weight gains and sequence of dental eruptions in infant owl monkeys (*Aotus trivirgatus*). In *Nursery care of nonhuman primates* (ed. G.C. Ruppenthal) pp. 321–8. Plenum Press, New York.

Hladik, C.M., Hladick, A., Bousset, J., Valdebouze, P., Viroben, G., and Delort-Laval, J. (1971). Le régime alimentaire des primates de L'Ile de Barro Colorado (Panama). *Folia Primatologica*, **16**, 85–122.

Hunter, J., Martin, R.D., Dixson, A.F., and Rudder, B.C. (1979). Gestation and interbirth intervals in the owl monkey (*Aotus trivirgatus griseimembra*). *Folia Primatologica*, **31**, 165–75.

ISIS (1989). *Species distribution report abstract: mammals.* ISIS, Minnesota.

IUCN (1990). *1990 IUCN red list of threatened animals.* IUCN, Gland, Switzerland.

Johnsen, D.O. and Whitehair, L.A. (1986). Research facility breeding. In *Primates. The road to self-sustaining populations* (ed. K. Benirschke), pp. 499–511. Springer-Verlag, New York.

Johnson, L.D., Petto, A.J., Boy, D.S., Sehgal, P.K., and Beland, M.E. (1986). The effect of perinatal and juvenile mortality on colony-born production at the New England Regional Primate Research Center. In *Primates. The road to self-sustaining populations* (ed. K. Benirschke), pp. 771–9. Springer-Verlag, New York.

Jones, M.L. (1986). Successes and failures of captive breeding. In *Primates. The road to self-sustaining populations* (ed. K. Benirschke), pp. 251–60. Springer-Verlag, New York.

King, N.W. Jr. and Chalifoux, L.V. (1986). Prenatal and neonatal pathology of captive nonhuman primates. In *Primates. The road to self-sustaining populations* (ed. K. Benirschke), pp. 763–70. Springer-Verlag, New York.

Ma, N.S.F., Jones, T.C., Miller, A.C., Morgan, L.M., and Adams, E.A. (1976). Chromosome polymorphism and banding patters in owl monkeys (*Aotus*). *Laboratory Animal Science*, **26**, 1022–36.

Melendez, L.V., Espana, C., Hunt, R.D., Daniel, M.D., and Garcia, P.G. (1969). Natural *Herpes simplex* infection in the owl monkey (*Aotus trivirgatus*). *Laboratory Animal Care*, **19**, 38–45.

Meritt, D.A. Jr. (1977). The owl monkey *Aotus trivirgatus*: husbandry, breeding and behaviour. *Proceedings of the American Association of Zoo Parks and Aquariums*. 23–4.

Meritt, D.A. Jr. (1980). Captive reproduction and husbandry of the douroucouli (*Aotus trivirgatus*) and the Titi monkey (*Callicebus spp*) *International Zoo Yearbook*, **20**, 52–9.

Meritt, G.F. and Meritt, D.A. Jr. (1978). Hand-rearing techniques for douroucoulis (*Aotu strivirgatus*). *International Zoo Yearbook*, **18**, 201–4.

Rouse, R., Bronson, R.T., and Sehgal, P.K. (1981). A retrospective study of etiological factors of abortion in the owl monkey *Aotus trivirgatus*. *Journal of Medical Primatology*, **10**, 199–204.

Sehgal, P.K., Bronson, R.T., Brady, P.S., McIntyre, K.W., and Elliot, M.W. (1980). Therapeutic efficacy of vitamin E and selenium in treating hemolytic anaemia of owl monkeys (*Aotus trivirgatus*). *Laboratory Animal Science*, **30**, 92–8.

Wolfheim, J.H. (1983). *Primates of the world*, pp. 237–45. University of Washington Press, Seattle.

Wright, P.C. (1981). The night monkeys Genus *Aotus*. In *Ecology and behaviour of neotropical primates*, Vol. 1 (ed. A.F. Coimbra-Filho and R.A. Mittermeier), pp. 211–40. Academia Brasiliera de Ciencias, Rio de Janeiro.

Squirrel monkey

10 Squirrel monkey

Species
The squirrel monkey   *Saimiri sciureus*

ISIS No.   1406006008002

Status, subspecies, and distribution
Thorington (1985) classified animals in the genus Saimiri into two species; *Saimiri sciureus*, of which he proposed there are four subspecies: *S.s. sciureus*, *S.s. cassiguarensis*, *S.s. boliviensis*, and *S.s. oerstedii*, and *S. madeirae*. However, research on the taxonomy of Saimiri is continuing (for example, Van de Berg *et al.* 1987). These animals are found in wet and dry tropical forests (Baldwin 1985) of the Amazon Basin and surrounding lands (Brazil, Peru, Bolivia, Colombia, and Guyana), but there is also a population in Panama and Costa Rica (Thorington 1985). The latter, which are *S.s. oerstedii*, are listed as endangered (Mittermeier 1986; IUCN 1990).

Johnsen and Whitehair (1986) estimated the number of actual and potential breeders in US principal research breeding centres in 1984 to be 2248, and the number of live births that year were 293. Squirrel monkeys are quite widely kept in zoos and a census of the population among ISIS members in 1984 indicated a population of 409 (Flesness 1986).

Sex ratio
In an analysis of 512 births, Rasmussen *et al.* (1980) found that the sex ratio was 1.3 males to 1 female. In a smaller sample (86 births), Clewe (1969) found a ratio of 1.6 male to 1 female.

Social structure
In the wild, squirrel monkeys live in groups ranging in size from 10 to several hundred animals (Baldwin 1985). Typical groups include 65 per cent infants and subadults, 29 per cent adult females, and 6 per cent adult males (Baldwin 1985). All-male subgroups are sometimes seen. In the breeding season males fight, sometimes fiercely, and unstable dominance hierarchies are formed. However, copulation does not appear to be clearly related to rank in the hierarchy.

In one captive colony, breeding groups were up to 30 females with two males, or 12 to 18 females with one male (Rasmussen *et al.* 1980). Dukelow (1985) in a review of breeding colony management reported that typical breeding group sizes were 10 to 15 females with 2 to 3 males (for example, Kaplan 1977) or, in smaller cages, 3 females to 1 male.

Breeding age
Wild females become sexually mature at about 2.5 years of age, but males remain in a subadult state for 2 to 3 years beyond this (Baldwin 1985). Taub (1980) found that 22 per cent of all animals in a captive colony became pregnant for the first time between 2.5 and 3 years of age, and a further 40 per cent became pregnant for the first time between 3.5 and 4 years.

Longevity
Jones (1980) recorded an individual still alive at 12.3 years, but it is likely that maximum lifespan exceeds this considerably. Kaplan (1977) reported that one female in his colony produced viable babies for nine consecutive years.

Seasonality
The squirrel monkey has a clear seasonal reproductive cycle and births in the wild generally coincide with the wet season (Baldwin 1985). In the captive colony described by Rasmussen *et al.* (1980) births were recorded in every month of the year but 72 per cent occurred in May, June, July, and August. The environmental cues involved in this clear seasonality are not known (Dukelow 1985). Although they are fertile throughout the year (Dukelow 1985), males show

a seasonal variation in weight. In the mating season, they gain 85–222 g and take on a 'fatted' appearance, with increased bulk in the shoulders (Baldwin 1985). Testis size also varies, from an average of 730 mg outside the breeding season to 1300 mg during it (Dumond and Hutchinson 1967).

Clewe (1969) allowed males access to groups of females for 30 minutes a day during the mating season, and observed that some females were mated daily for up to nine weeks, and that mating also occurred during the first four months of pregnancy, and also during lactation. The oestrous cycle is 7 to 12 days (Dukelow 1985).

Gestation

Dukelow (1985) considered that the evidence from studies in which sophisticated techniques had been used to pinpoint conception dates, indicated a normal gestation period of 145 to 155 days. Kerber et al. (1977) found a mean of 147 days in 10 females of Colombian/Brazilian origin. Some previous studies indicated means of around 165 days (for example, Delort et al. 1976).

Pregnancy diagnosis

Abdominal distension may be apparent from 75 to 100 days of pregnancy (Hopf 1967), and uterine enlargement can be detected by palpation from 42 to 56 days of pregnancy (Rosenblum 1968; Kaplan 1977). Pregnancy can also be diagnosed quite reliably by detection of chorionic gonadotrophin in blood or urine; and a urine test kit has been developed (Hodgen et al. 1978).

Birth

Births usually occur at night (Hopf 1967; Clewe 1969; Kaplan 1977), the majority between 02.00 h and 06.00 h (Manocha and Long 1977); and the time to delivery, from the onset of visible signs of labour, has been recorded as 88 to 165 minutes (Takeshita 1962; Hopf 1967), although breech presentations may take longer (Bowden et al. 1967). Females showing signs of parturition during the day are likely to be having problems (Kaplan 1977).

Clewe (1969) observed that there was a marked increase in water consumption in the fifth month of pregnancy. Water consumption fell to the non-pregnant level 2 to 16 days prior to parturition, and this gave some clue that parturition was imminent.

Litter size

The litter size is 1. We have found only one record of twins and both were born dead (Kaplan 1977).

Adult weight

Adult weights generally fall within the ranges 700 to 1100 g for males and 480 to 750 g for females (Bowden et al. 1967; Long and Cooper 1968; Middleton and Rosal 1972; Kaack et al. 1979), with means of about 950 g for males and 650 g for females.

Neonate weight

In their analysis of 459 births, Rasmussen et al. (1980) found that mean birth weight of 43 aborted or stillborn was 89 g, that of 41 liveborn that died within a week was 101 g, and that of 375 that survived at least one week was 111 g. These authors also showed that the mean weight of 237 liveborn males (112 g) was greater than the mean weight of the female babies (106 g). In their colony mean birth weight varied significantly from year to year, from 123 g, ($n = 16$) in 1969, to 102 ($n = 64$) in 1975. There was no correlation between maternal age and birth weight, but birth weight was significantly positively correlated with maternal weight (Rasmussen et al. 1980). There may be variation in both adult and neonate weight between subspecies. Kaplan (1977) found that Colombian infants were consistently heavier than Peruvian infants and Bolivian infants appeared to be intermediate.

Adult diet

Wild squirrel monkeys feed largely on fruit and insects but leaves have also been found in the stomach contents (Thorington 1968; Izawa 1975; Mittermeier and Van Roosmalen 1981).

Diets offered in captivity commonly consist of proprietary primate chow supplemented with fresh fruit, and small amounts of cheese, crickets, and hard-boiled eggs. Meat should be used with caution, in our experience, as this species is particularly susceptible to acute disease caused by

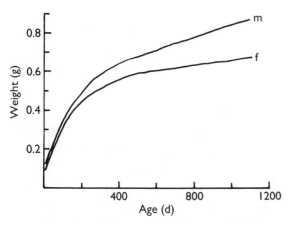

Fig. 10.1. Average growth curve of nursery-reared male (m) and female (f) squirrel monkeys. From Russo et al. (1980). (These curves are similar to those reported by Long and Cooper 1968, for mother-reared animals.)

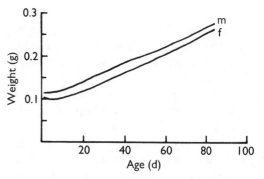

Fig. 10.2. Typical neonatal growth of nursery-reared male (m) and female (f) squirrel monkeys. From Ausman et al. (1970), Hinkle and Session (1972), and Kaack et al. (1979).

Toxoplasma and the animals may acquire the infection from meat.

Vitamin D_3 is essential and Lehner et al. (1967) found that 1250 iu/kg of dry diet was sufficent to prevent bone lesions. Dietary vitamin C is also required and fresh fruit is a suitable source (Lang 1968). The nutrition of the squirrel monkey has received considerable study (Ausman et al. 1985).

Adult energy requirements

Although the nutrition of infant squirrel monkeys has received considerable attention, we are unaware of data on the energy requirements of adults.

Growth

Squirrel monkeys continue to gain weight until at least 3 years old, although the rate of growth in 2 and 3 year olds is much lower than that during the first year (Fig. 10.1). Nursery-reared infants may lose weight during the first week (Fig. 10.2) but then between 10 and 100 days of age they grow at about 2.5 to 3.5 g/d (Figs. 10.1 and 10.2). After 1.5 to 2 years of age, weight gain averages about 0.3 g/d in males and 0.1 g/d in females (Fig. 10.1). There is considerable variation in the pattern of growth between individuals: the males reared by Russo et al. (1980) ranged from about 490 to 720 g at 1 year of age, and the females ranged from about 400 to 670 g. The weight gain of surrogate-reared infants reared by Kaplan (1979) was very similar to that of mother-reared infants in his colony. His results indicated that Columbian infants tended to grow more rapidly than those from Peru, and the growth rate of Bolivian infants was intermediate.

Milk and milk intake

Buss and Cooper (1972) collected 13 milk samples from 4 squirrel monkeys between 93 and 153 days of lactation, and found that the composition was, on average, 5.1 per cent fat, and 3.5 per cent crude protein, 6.3 per cent carbohydrate (lactose), 0.3 per cent ash, and 84.8 per cent water. From this it appears that the fat, protein, carbohydrate, and ash composition of the dry matter was 0.34, 0.23, 0.41, and 0.02 respectively. These authors also analysed the fatty acid, amino acid, and major mineral composition of their milk samples.

Squirrel monkeys have been successfuly reared on human milk replacers (Kaplan 1977), such as SMA (Kaack et al. 1979) and Similac with Iron (Ausman et al. 1970). Others have made up their own formulae from milk replacers and dairy products, with added vitamin D_3, to try to mimic squirrel monkey milk composition more closely (see Hinkle and Session 1972).

Ausman et al. (1985) pointed out that the use of human milk replacers (which have lower protein and fat levels than squirrel monkey milk) could affect the growth rate of infants, and cited unpublished data which indicate that low fat, high lactose milks (and here they were presumably referring to human milk replacers) were associated with poor neonatal survival. Furthermore, they refer to other unpublished data which indicated that, in animals a few weeks old, rate of weight gain was greater when fed diets of 1 kcal/ml than when fed diets of 0.7 kcal/ml (the typical energy density of human milk replacers). Ausman et al. (1985) also mention that infant squirrel monkeys fed solely on human milk replacers until 12–16 weeks old, developed a yellow hair coat with no black stripes. The cause of this abnormal pigmentation was unknown. Although human milk replacers are easy to obtain, of defined composition, and able to support growth, there is therefore evidence that they are not ideal. Offering solid foods at any early age (35 days or before) may enable infants to avoid nutrient imbalances when fed on human milk replacers.

Based on their studies of infant squirrel monkeys, Ausman et al. (1985) devised recipes for milk replacers for infants in their first 2 months and thereafter. The reader is referred to their paper for details. Primilac is probably a more suitable off-the-shelf replacer than those designed for human babies and its protein, fat, and carbohydrate levels are like those in the do-it-yourself formula described by Ausman et al. (1985). However, controlled comparisons have not, as far as we are aware, been undertaken.

Some have reported feeding half-strength milk replacers for the first two days (Ausman et al. 1970). There is a case for the very first feeds being of electrolyte solution to reduce the risk of aspiration pneumonia.

Anderson (1986) suggested that the volume of milk provided at each feed during the first week should be 1–3 ml. The animals reared by Hinkle and Session (1972) showed a rapid increase in milk intake from about 25 ml/d at 7 days old to a plateau of about 11 ml/d from 45 days onwards (Fig. 10.3). Their formula had an energy density of 0.9 kcal/ml, and it appears from their data that the daily metabolizable energy intake increased from about 125 kcal/kg$^{0.75}$ when 10 days old to 300 kcal/kg$^{0.75}$ when 35 days old (Fig. 10.4). The

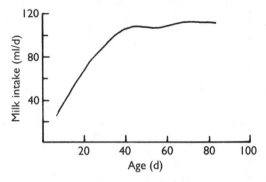

Fig. 10.3. Approximate daily milk intake of nursery-reared squirrel monkeys. From Hinkle and Session (1972).

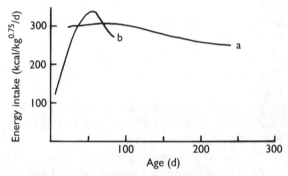

Fig. 10.4. Approximate daily energy intake provided by milk in hand-reared squirrel monkeys per metabolic weight in relation to age: (a) estimated from data of Russo et al. (1980) and Ausman et al. (1985); (b) estimated from data of Hinkle and Session (1972).

calorie requirements in relation to body weight listed by Ausman et al. (1985) are consistent with a daily intake of about 300 kcal/kg$^{0.75}$ between 25 and 100 days of age, followed by a gradual decline to 250 kcal/kg$^{0.75}$ by 250 days of age.

Lactation and weaning

The infant finds its way to the nipples very soon after birth (Takeshita 1962; Hopf 1967; Rosenblum 1968). Anderson (1986) suggested that squirrel monkeys may begin to eat solid food at 4 to 20 days of age. The babies reared by Kaack et al. (1979) were offered solid food from 35 days of age and were completely weaned by 38 weeks and those whose rearing was described by Kaplan

(1977) were weaned by 6 months of age. Buss and Cooper (1972) noted that mother-reared squirrel monkeys aged between 3 and 5 months were eating substantial amounts of solid food. By 5 to 6 months of age, the mother-reared infants are completely weaned and mothers actively discourage contact (Coe et al. 1985; see also Rosenblum 1968).

Feeding

In the management system described by Hinkle and Session (1972) babies were fed every 2 hours between 06.30 h and 22.30 h for the first 7 days after birth, every 4 hours for the next 2 weeks, and thereafter the animals fed themselves *ad libitum* from a bottle attached to their surrogate mothers. Ausman et al. (1970) began hand-feeding babies 4 to 6 hours after birth, at 2 hour intervals (not during the night), using a 30 ml bottle with a small nipple. Their animals began self-feeding at about 14 days old. From about 3 to 4 weeks of age, *ad libitum* water was also available.

Accommodation

Ausman et al. (1970) measured rectal temperature in nursery-reared squirrel monkeys housed in incubators at 29.4 °C, and found that it increased from an average of 35.2 °C during the first week to 37.3 °C in infants between 15 and 21 days of age (see also Physical development). In the early neonatal period, a higher incubator temperature of 32–35 °C may be more suitable especially for low birth weight infants. After 3 or 4 weeks the infants reared by Ausman et al. (1970) were transferred to individual cages with plexiglass doors, in a room with an ambient temperature of 22 °C, and heating pads were provided when necessary to maintain body temperature. Reduction in environmental temperature should be gradual.

Infant management notes

Hinkle and Session (1972) kept newborn squirrel monkeys in incubators at 32.2 °C and gradually reduced the temperature, by 1.1 °C per week, until at 8 weeks the babies were transferred to cages in a room at 25 °C.

Various surrogate mothers for squirrel monkeys have been described. Hinkle and Session (1972) made disposable surrogates out of towel rolled into cylinders (20 cm long, 7.5 cm diameter), fixed with tape. These, and the towels lining the incubators, were replaced daily. A more sophisticated surrogate with a bottle and nipple enabling the infant to feed itself was described by Kaplan and Russel (1973); Kaplan (1973), and further modifications to this design were reported by Herzog and Hopf (1985). The surrogate made by the latter authors had an extra imperforate nipple, arm-like structures to help the infant pinpoint the site of the nipples, and a lever that when pressed squeezed the milk bottle and expressed a small drop of milk to assist inexperienced infants in learning to self-feed. It also had an electric rocking mechanism which rocked the surrogate through 6 degrees, 0.5 to 3 times a second. The latter made the infant relax and sleep. The rationale for the extra imperforate nipple was to prevent the development of thumb-sucking by providing something else for the baby to suck.

Physical development

Long and Cooper (1968) studied the sequence and age of eruption of the teeth of squirrel monkeys. The average age at eruption of the deciduous teeth was: upper and lower first incisors, 1.3 weeks; lower second incisors, 1.4 weeks; upper second incisors, 2.2 weeks; upper and lower canines, 3.4 weeks; lower first premolars, 4.8 weeks; upper first premolars, 4.9 weeks; lower second premolars, 5.7 weeks; upper second premolars, 6.1 weeks; lower third premolars, 8.3 weeks; and upper third premolars, 8.9 weeks. About three months later the permanent teeth begin to appear at the following average ages: lower first molars, 5.1 months; upper first molars, 5.5 months; lower second molars 7.0 months; upper second molars, 8.4 months; lower first incisors, 9.2 months; lower second incisors, 9.6 months; upper first incisors, 9.7 months; upper second incisors, 12.0 months; lower third molars, 12.3 months; lower third premolars, 12.8 months; upper second and third premolars, 13.0 months; lower first premolars, 14.3 months; upper first premolars, 14.7 months; lower second premolars, 15 months; upper third molars,

20.0 months; lower canines 20.5 months; and upper canines, 21.5 months.

It appears that full thermoregulatory ability takes several weeks, if not months, to develop. Ausman et al. (1970) observed that body temperature of nursery-reared infants kept at 29.4 °C averaged 35.2 °C during the first week and increased by 2 °C during the first three weeks. Coe et al. (1985) found that even at 14 weeks of age infant body temperature fell to 35.5 °C after 4 hours of forced maternal separation (they did not mention the ambient temperature). These results emphasize the need for the provision of higher environmental temperatures than necessary for adults (or heating pads to provide supplementary warmth), for nursery-reared infants until 4 to 5 months of age.

Behavioural development

Babies cling to their mothers' backs from birth. During the first week they begin to look around and, when 2 weeks old begin to reach down with their hands to manipulate nearby objects. They begin to make their first independent ventures when 3 to 5 weeks old, and by 8 weeks old, spend long periods off their mothers (Rosenblum 1968). After 4 months of age the infant is rarely carried by its mother during the day but may sleep on her back until 14 months old (Rosenblum 1968). Although lactation probably ceases by 6 months postpartum, infants may continue to show occasional oral contact with the mother's nipples until 10 or more months old (Rosenblum 1968).

Between 5 and 20 weeks of age infants often develop a relationship with an 'aunt' (typically a nulliparous associate of its mother) and may be carried by her for extended periods (Rosenblum 1968). Infants begin to show social play (chasing and wrestling) when 2 months old (Rosenblum 1968). It would therefore seem appropriate that nursery-reared infants should be group-housed from this age or before, at least for part of the day. Kaplan (1977) reported that infants reared alone or with only a surrogate mother for four months or more may have great difficulty when introduced to social groups.

Surrogate-reared squirrel monkeys may develop atypical behaviours; most commonly thumb, genital, and tail-sucking (Kaplan 1973). However, the incidence varies between colonies and may depend upon the quality of the environment and especially on the design of the surrogate (Hennessy 1985). Nonetheless, squirrel monkeys appear to be more resilient, in terms of subsequent behaviour normality, to artificial rearing than macaques. Hennessy (1985) found no significant difference in survival to 1 week of age of first offspring of surrogate-reared and mother-reared females (but see Disease and mortality).

Disease and mortality

Abortion, still birth, and neonatal mortality rates have been found to be high in captive squirrel monkeys. Johnson et al. (1986) reported a lower survival to 2 years of age (5 out of 17 born) in this species than any of the other 7 species reared at their center in 1983. These authors, in a review of colony statistics 1967 to 1983, found that 169 (41 per cent) of the 414 births were stillborn, 65 (27 per cent) of 245 liveborn died within the first 30 days, and a further 66 died before reaching 2 years of age. A high incidence of still births was also found by Aksel and Abee (1983); out of 40 pregnancies, 6 were aborted, and 123 were born dead at full term. Hopf (1981) reported that of 49 viable colony-born infants, 13 died and 2 were rejected then surrogate-reared.

The outcome of 512 pregnancies was reviewed by Rasmussen et al. (1980). Of these, 430 (84 per cent) resulted in a live birth, 61 (12 per cent) were stillborn, and 21 (4 per cent) were aborted. Of the liveborn 17 (4 per cent) died within 24 hours and another 30 (7 per cent) died between 2 and 7 days of age. Of 124 live births recorded by Padovan and Cantrell (1983), 38 (31 per cent) died prior to 12 months of age. Kaplan (1977) recorded a comparatively low rate (15 per cent) of mortality to 6 months of age in mother-reared babies and found that 80 per cent of the deaths in this age group occurred between 4 and 8 weeks of age. He speculated that waning maternal antibody levels might have rendered the 4-week-old babies susceptible to infections. Williams and Abee (1986) reported an overall mean neonatal mortality rate of 43.6 per cent in their colony.

High rates of abortion and infant mortality may be due to many factors but a few, probably major, causes have been suggested. Kaplan (1977) found that a large proportion of resorptions and abortions occurred in females stressed by being handled. Manocha and Long (1977) showed that there was a marked difference in abortion rate between females kept on low- and high-protein diets. In their colony, the total abortion rate in females fed diets in which protein provided only 8 per cent of the calories, was 50 per cent (65 out of 131 pregnancies), whilst that in females fed diets where protein provided 21 per cent of the calories was 8 per cent (7 out of 87 pregnancies). Taub et al. (1978) found that reproductive success was not related to age but that colony-born females deprived of sexual and maternal experience during development had higher rates of fetal and neonatal deaths than feral females (see also Lehner 1985).

Aksel and Abee (1983) found that there was a significant difference in the diameter of the pelvic outlet (the distance between the inferior lateral margins of the obturator foramen of the ischial rami) between those females that had live and those that had stillborn babies. Radiographic measurement of the diameter of the pelvic outlet appeared to provide a good predictor of the chances of live birth. These authors suggested that a contracted pelvic outlet might interfere with safe delivery, but did not comment about the possible causes of this apparently important variation in the size of the pelvic outlet. It seems unlikely that fetal–pelvic disproportion is a common problem in wild squirrel monkeys.

The results presented by Taub (1980) suggest that there was no marked difference in success rate between first and second pregnancies. Of 60 first pregnancies, 23 (38 per cent) resulted in fetal death and 19 (50 per cent) of the 37 liveborn died before 6 months of age. The results for second pregnancies were similar: of 39 pregnancies, 12 (31 per cent) resulted in fetal death, and of the liveborn, 14 (52 per cent) died before 6 months of age. He also found that those females that gave birth to live young at their first pregnancy were more likely to have live young at their second, than those whose first babies had been born dead.

As mentioned in the section on neonate weight, weight at birth has a clear effect on neonatal viability. In one colony, supplementation of the diet with folic acid had a significant effect on birth weight (Rasmussen et al. 1980).

Greater survival rates to 6 months of age have been reported in artificially reared squirrel monkeys than in those reared by their mothers (Kaplan 1977). It appears from the results presented in Kaplan (1979) that in his colony, mortality to 6 months of age was 23 per cent (32 out of 136) among mother-reared infants but only 7 per cent (3 out of 41) of those reared artificially.

King and Chalifoux (1986), in a review of the pathology of 189 squirrel monkeys which died during the neonatal period (the first 30 days), found no lesions in 127 (67 per cent) of these, but 18 (10 per cent) died as a result of trauma at birth or dystocia, and a further 18 (10 per cent) died because of mutilation by the parents. A wide miscellany of other conditions were reported. These authors considered that parental neglect accounted for many of the deaths in which no specific lesions were found, and that hypothermia and dehydration may lead to the deaths of abandoned babies quite rapidly. Rasmussen et al. (1980) considered that early neonatal mortality was largely due to accidents, premature birth, and trauma (injury during birth or by cage mates). Padovan and Cantrell (1983) also found that trauma was an important cause of early infant mortality (fracture of the skull can occur when the infant's head is knocked against cage furniture or walls whilst riding on its mother), and that bronchopneumonia was a common cause of death in older infants.

Congenital defects are uncommon but oral–facial defects have been reported (Hoopes and Jerome 1987).

Preventative medicine

There are large differences in rates of abortion and infant mortality between colonies (see Disease and mortality), which suggests that these are related to colony husbandry. Breeding group composition, housing, nutrition, and management practices probably all influence colony breeding and rearing success, and high standards are important. Although normal behavioural development seems to be more robust in the squirrel monkey than in the rhesus macaque, animals reared in isolation or with inadequate socialization are unlikely to be capable of rearing infants.

Squirrel monkeys are susceptible to a wide range of infectious diseases but routine vaccinations are not carried out in most colonies at present.

Indications for hand-rearing

Small or premature infants have a very low neonatal survival rate and their chances of survival may be increased by hand-rearing.

Reintegration

Kaplan (1977) considered that infants should probably be housed together in pairs or in small groups from as young an age as is practical to facilitate normal social development. Introductions to a new group should be made gradually. The introduction of adults to established groups is associated with aggression (Williams and Abee 1988), as in many primate species.

Various methods of reintroducing nursery-reared squirrel monkeys to the colony have been explored by Ricker et al. (1985). They placed 4-to-6-week-old infants in peer groups of 2-4 animals, and then gradually integrated them into mixed age and sex groups when 12-15 weeks old.

References

Aksel, S. and Abee, C.R. (1983). A pelvimetry method for predicting perinatal mortality in pregnant squirrel monkeys (*Saimiri sciureus*). *Laboratory Animal Science*, 33, 165-7.

Anderson, J.H. (1986). Rearing and intensive care of neonatal and infant nonhuman primates. In *Primates. The road to self-sustaining populations* (ed. K. Benirschke), pp. 747-62. Springer-Verlag, New York.

Ausman, L.M., Hayes K.C., Lage, A., and Hegsted, D.M. (1970). Nursery care and growth of Old and New World monkeys. *Laboratory Animal Care*, 20, 907-13.

Ausman, L.M., Gallina, D.L. and Nicolosi, R.J. (1985). Nutrition and metabolism of the squirrel monkey. In *Handbook of squirrel monkey research* (ed. L.A. Rosenblum and C.L. Coe), pp. 349-78. Plenum Press, New York.

Baldwin, J.D. (1985). The behaviour of squirrel monkeys (Saimiri) in natural environments. In *Handbook of squirrel monkey research* (ed. L.A. Rosenblum, and C.L. Coe), pp. 35-53. Plenum Press, New York.

Bowden, D., Winter, P., and Ploog, D. (1967). Pregnancy and delivery behaviour in the squirrel monkey and other species. *Folia Primatolgica*, 5, 1-42.

Buss, D.H. and Cooper, R.W. (1972). Composition of squirrel monkey milk. *Folia Primatologica*, 17, 285-91.

Clewe, T.H. (1969). Observations on reproduction of squirrel monkeys in captivity. *Journal of Reproduction and Fertility Supplement*, 6, 151-6.

Coe, C.L., Wiener, S.G., Rosenberg, L.T., and Levine, S. (1985). Physiological consequences of maternal separation and loss in the squirrel monkey. In *Handbook of squirrel monkey research* (ed. L.A. Rosenblum and C.L. Coe), pp. 127-48. Plenum Press, New York. 148

Delort, D., Chopard, M., and Tachon, J. (1976). The management of a squirrel monkey (*Saimiri sciureus*) unit. *Laboratory Animal Science*, 26, 301-4.

Dukelow, W.R. (1985). Reproductive cyclicity and breeding in the squirrel monkey. In *Handbook of squirrel monkey research* (ed. L.A. Rosenblum and C.L. Coe), pp. 169-90. Plenum Press, New York.

Dumond, F.W. and Hutchinson, T.C. (1967). Squirrel monkey reproduction: the 'fatted' male phenomenon and seasonal spermatogenesis. *Science*, 1467-70.

Flesness, N.R. (1986). Captive status and genetic considerations. In *Primates. The road to self-sustaining populations* (ed. K. Benirschke,), pp. 845-56. Springer-Verlag, New York. 856

Hennessy, M.B. (1985). Effects of surrogate rearing on the infant squirrel monkey. In *Handbook of squirrel monkey research* (ed. L.A. Rosenblum and C.L. Coe), pp. 149-68. Plenum Press, New York.

Herzog, M. and Hopf, S. (1985). Some improvements of mother surrogates for squirrel monkeys: a technical note. *Laboratory Primate Newsletter*, 24, 1-2.

Hinkle, D.K. and Session, H.L. (1972). A method for hand rearing of *Saimiri sciureus*. *Laboratory Animal Science*, 22, 207-9.

Hodgen, G.D., Stolzenberg, S.J., Jones, D.C.L., Hildebrand, D.F., and Turner, C.K. (1978). Pregnancy diagnosis in squirrel monkeys: Hemagglutination test, radioimmunoassay, and bioassay of chorionic gonadotrophin. *Journal of Medical Primatology*, 7, 59-64.

Hoopes, C.W and Jerome, C.P. (1987). Oral-facial clefts and associated malformations in the squirrel (*Saimiri sciureus*). *Journal of Medical Primatology*, 16, 203-9.

Hopf, S. (1967). Notes on pregnancy, delivery and infant survival in captive squirrel monkeys. *Primates*, 8, 323-32.

Hopf, S. (1981). Conditions of failure and recovery of maternal behaviour in captive squirrel monkeys (Saimiri). *International Journal of Primatology*, 2, 335-49.

IUCN (1990). *1990 IUCN red list of threatened animals*. pp. 10. IUCN, Gland, Switzerland.

Izawa, K. (1975). Foods and feeding behaviour of monkeys of the upper Amazon basin. *Primates*, 17, 367-99.

Johnsen, D.O. and Whitehair, L.A. (1986). Research facility

breeding. In *Primates. The road to self-sustaining populations* (ed. K. Benirschke), pp. 499–511. Springer-Verlag, New York.

Johnson, L.D., Petto, A.J., Boy, D.S., Sehgal, P.K., and Beland, M.E. (1986). The effect of perinatal and juvenile mortality on colony-born production at the New England Regional Primate Research Center. In *Primates. The road to self-sustaining populations* (ed. K. Benirschke), pp. 771–9. Springer-Verlag, New York.

Jones, M.L. (1980). Lifespan in mammals. In *The comparative pathology of zoo animals* (ed. R.J. Montali, and G. Migaki), pp. 495–509. Smithsonian Institution Press, Washington D.C.

Kaack, B., Brizzee, K.R., and Walker, L. (1979). Some biochemical blood parameters in the developing squirrel monkey. *Folia Primatologica*, 32, 309–17.

Kaplan, J.N. (1973). Growth and behaviour of surrogate-reared squirrel monkeys. *Developmental Psychobiology*, 7, 7–13.

Kaplan, J.N. (1977). Breeding and rearing squirrel monkeys *Saimiri sciureus*) in captivity. *Laboratory Animal Science*, 27, 557–67.

Kaplan, J.N. (1979). Growth and development of infant squirrel monkeys during the first six months of life. In *Nursery care of nonhuman primates* (ed. G.C. Ruppenthal), pp. 153–64. Plenum Press, New York.

Kaplan, J. and Russel, M. (1973). A surrogate for rearing infant squirrel monkeys. *Behaviour Research Methods and Instrumentation*, 5, 379–80.

Kerber, W.T., Conaway, C.H., and Moore Smith, D. (1977). The duration of gestation in the squirrel monkey (*Saimiri sciureus*). *Laboratory Animal Science*, 27, 700–2.

King, N.W. and Chalifoux, L.V. (1986). Prenatal and neonatal pathology of captive nonhuman primates. In *Primates. The road to self-sustaining populations* (ed. K. Benirschke), pp. 763–70. Springer-Verlag, New York.

Lang, C.M. (1968). The laboratory care and clinical management of Saimiri (squirrel monkeys). In *The squirrel monkey* (ed. L.A. Rosenblum, and R.W. Cooper), pp. 393–416. Academic Press, New York.

Lehner, N.D.M. (1985). Effects of rearing practices on performance of squirrel monkeys. *Laboratory Animal Science*, 35, 529.

Lehner, N.D.M., Bullock, B.C., and Clarkson, T.B. (1967). Biological activities of vitamins D2 and D3 for growing squirrel monkeys. *Laboratory Animal Care*, 17, 483–93.

Long, J.O. and Cooper, R.W. (1968). Physical growth and dental eruption in captive-bred squirrel monkeys, *Saimiri sciureus* (Leticia, Columbia). In (ed. L.A. Rosenblum and R.W. Cooper), pp. 193–205. *The squirrel monkey*. Academic Press, New York.

Manocha, S.L. and Long, J. (1977). Experimental protein malnutrition during gestation and breeding performance of squirrel monkeys, *Saimiri sciureus*. *Primates*, 18, 923–30.

Middleton, C.C. and Rosal, J. (1972). Weights and measurements of normal squirrel monkeys (*Saimiri sciureus*). *Laboratory Animal Science*, 22, 583–6.

Mittermeier, R.A. (1986). Primate conservation priorities in the Neotropical region. In *Primates. The road to self-sustaining populations* (ed. K. Benirschke), pp. 221–40. Springer-Verlag, New York.

Mittermeier, R.A. and Van Roosmalen, G.M. (1981). Preliminary observations of habitat utilisation and diet in eight Surinam monkeys. *Folia Primatologica*, 36, 1–39.

Padovan, D. and Cantrell, C. (1983). Causes of death in infant rhesus and squirrel monkeys. *Journal of the American Veterinary Medicine Association*, 183, 1182–4.

Rasmussen, K.M, Ausman, L.M., and Hayes K.C. (1980). Vital statistics from a laboratory breeding colony of squirrel monkeys (*Saimiri sciureus*). *Laboratory Animal Science*, 30, 99–106.

Ricker, R.B., Brady, A.C., and Abee, C.R. (1985). Nursery rearing strategies for captive-born squirrel monkeys. *Laboratory Animal Science*, 35, 529.

Rosenblum, L.A. (1968). Mother–infant relations and early behavioural development in the squirrel monkey. In *The squirrel monkey* (ed. L.A. Rosenblum and R.W. Cooper), pp. 207–33. Academic Press, New York.

Russo, A.R., Ausman, L.M., Gallina, D.L., and Hegsted, D.M. (1980). Developmental body composition of the squirrel monkey (*Saimiri sciureus*). *Growth*, 44, 271–86.

Takeshita, H. (1962). On the delivery behaviour of squirrel monkeys (*Saimiri sciureus*) and a mona monkey (*Cercopithecus mona*). *Primates*, 3, 59–72.

Taub, D.M. (1980). Age at first pregnancy and reproductive outcome among colony-born squirrel monkeys (*Saimiri sciureus, Brazilian*). *Folia Primatologica*, 33, 262–72.

Taub, D.M., Adams, M.R., and Auerbach, K.G. (1978). Reproductive performance in a breeding colony of Brazilian squirrel monkeys (*Saimiri sciureus*). *Laboratory Animal Science*, 28, 562–6.

Thorington, R.W. Jr. (1968). Observations of squirrel monkeys in a Colombian forest. In *The squirrel monkey* (ed. L.A. Rosenblum and R.W. Cooper), pp. 69–87. Academic Press, London.

Thorington, R.W. Jr. (1985). The taxonomy and distribution of squirrel monkeys (*Saimiri*). In *Handbook of squirrel monkey research* (ed. L.A. Rosenblum and C.L. Coe), pp 1–33. Plenum Press, New York.

Van de Berg, J.L., Cheng, M.L., Moore, C.M., and Abee, C.R. (1987). Genetics of squirrel monkeys (Genus *Saimiri*): implications for taxonomy and research. *International Journal of Primatology*, 8, 423.

Williams, L.E. and Abee, C.R. (1986). Reproductive performance through four years in a laboratory housed colony of *Saimiri boliviensis*. *Primate Report*, 14, 119–20.

Williams, L.E. and Abee, C.R. (1988). Aggression with mixed age–sex groups of Bolivian squirrel monkeys following single animal introductions and new group formations. *Zoo Biology*, 7, 139–45.

Vervet monkey

11 Vervet monkey (Green monkey or grivet)

Species

The vervet monkey (green monkey or grivet) *Cercopithecus aethiops*

ISIS No.   1406008002001001

Status, subspecies, and distribution

The species is common and widely distributed in Africa south of the Sahara, and 21 subspecies have been recognized (Hall and Gartlan 1965; Gartlan 1969; Wolfheim 1983). It is a highly adaptable terrestrial or semi-terrestrial species (Mitchell 1979a; Rowell, 1984), found in savanna, woodland edges, rain forests, and urban parks (Bloomstrand and Maple 1987), and typically around riverine acacia trees (Rowell 1984). It does not occur in closed forests and arid areas (Gartlan 1969). It is also found on the West Indian islands of St. Kitts and Barbados to which it was introduced in the eighteenth century (Poirier 1972; Horrocks 1986).

In the past, these animals were quite widely used in biomedical research (Rowell 1970). A total of 21 779 were exported to the United States between 1968 and 1973, and 11 232 were exported to the United Kingdom between 1965 and 1975 (Wolfheim 1983). They are not recorded as important biomedical research species now (Johnsen and Whitehair 1986). Zoo populations appear to be small. According to the ISIS records of 30 June 1986, there were 59 *C. aethiops* of various subspecies held in captivity in zoos in the United States (Bloomstrand and Maple 1987).

Sex ratio

In the colony described by Kushner *et al.* (1982), 97 males and 89 females were born. This slight skew in the sex ratio is not statistically significant. Among free-living adult vervets, however, the sex ratio of males to females was reported to be 1:1.4 (Hall and Gartlan 1965) and 1:2 (Lancaster, 1971) in Africa, and 1:2 on St. Kitts (Poirier, 1972).

Social structure

The social structure of *C. aethiops* consists of stable multi-male (Mitchell 1979a; Rowell 1984; Bloomstrand and Maple 1987) and multi-female groups (Bloomstrand and Maple 1987; Bramblett and Coelho 1987). Immature animals may form up to 45 per cent of the group (Hall and Gartlan 1965). The males migrate from their natal groups (Mitchell 1979b; Bramblett and Coelho 1987), usually at the age of 4.5 to 5.5 years (Henzi and Lucas 1980). Occasionally females migrate from their natal groups taking all their offspring (Mitchell 1979b; Henzi and Lucas 1980). Estimates of mean group size vary from 12 to 30 members (Hall and Gartlan 1965; Poirier 1972; McGuire 1974), but groups of up to 174 animals have been observed (Galat and Galat-Luong 1978). Groups provided with extra food by mean tended to be larger than unprovisioned groups (average sizes of 28 and 16 respectively) (Lee *et al.* 1986).

Breeding age

Puberty, or menarche, in the female is reached at approximately 30 months of age (Bramblett *et al.* 1975) when the vaginal orifice is fully opened and the canalization of the vagina is complete (Hafez 1971). Rowell (1984) considered that females first conceive in the wild at about 5 years of age, but Bramblett and Coelho (1987) observed pregnancies in 3-year-olds. In captivity, conception can occur at 2.5 of age years (Rowell 1984). Two females were reported to have produced their first infants at 39 and 43 months of age respectively (Bramblett *et al.* 1975). Males reach puberty and begin spermatogenesis at about the same time as menarche in females, before they are fully-grown and socially mature. In the wild, they therefore tend not to breed until several years later (Rowell 1984).

Longevity

There are several records of *C. aethiops* living for over 20 years in captivity, and one is reported to have lived in the Cairo Zoo for 24 years (Jones 1968).

Seasonality

On St. Kitts, most births occur from late May to early July (Poirier 1972), but about 20 per cent occur in December and January (McGuire et al. 1974). In Africa, vervet troops do show seasonal breeding (Lancaster 1971) but the time of year at which births occur varies between neighbouring troops (Lee et al. 1986). Michael and Zumpe (1971) recorded that births in Amboseli, Kenya, occurred from October to March, whereas by Lake Victoria they occurred from April to September. Bramblett et al. (1975) reported seasonal and synchronized births within a troop but observed that births occur during every month of the year. The seasonality may be dependent on social rather than climatic factors (Gartlan 1969). In captivity, breeding may occur throughout the year (Mallinson 1970, 1971) although in a colony in Pennsylvania significantly more births occurred during the warmer months of May to September (Kushner et al. 1982).

The duration of the oestrus cycle has been measured at 25 to 46 days (Rowell 1970; Michael and Zumpe 1971; Mitchell 1979a) but is usually about 1 month (Bramblett et al. 1975). The mean duration was found by Johnson et al. (1973) to be 30.9 days. Menstruation is rarely overt (Gartlan 1969; Rowell 1970; Mitchell 1979a) but can be detected by examination of saline vaginal lavages or smears (Gartlan 1969; Rowell 1970; Bramblett et al. 1975). It lasts from one to seven days (Bramblett et al. 1975). No sexual skin swelling occurs during the cycle (Gartlan 1969; Mitchell 1979a). Mating occurs throughout the cycle (Rowell 1970).

After parturition, cycling is resumed within seven or eight months in 70 per cent of females (Asanov 1972) and conception may occur during lactation (Hendrickx and Giles Nelson 1971; Kushner et al. 1982). Interbirth intervals show wide individual variation (Asanov 1972), but Kushner et al. (1982) found the mean to be 357 days in their colony.

Cyclic variation in testes size has been observed in the field (Michael and Zumpe 1971).

Gestation

The gestation period ranges from 161 to 165 days (Rowell 1970; Bramblett et al. 1975) or 163 to 165 days (Kushner et al. 1982). The mean in the colony of Johnson et al. (1973) was 163 days.

Pregnancy diagnosis

Pregnancy can be diagnosed 25–30 days after conception by rectal palpation: the non-gravid uterus is about 0.5 cm wide and 2 cm in length, but 30 days after conception it is about 1.25 cm wide and 3 cm in length.

A high leucocyte count has been observed between the fifth and tenth week of pregnancy. Palpation or ultrasound scanning can be used for pregnancy diagnosis (Rowell 1970). The use of a urine test for detection of pregnancy has also been reported (Andelman et al. 1985).

Birth

Births occur at night or in the early morning (Gartlan 1969; Mallinson 1971). In one delivery (Shively and Mitchell 1986) imminent labour was signalled by the behaviour of the female who was stretching and pulling on her cage. The female then began to grunt and to touch the vaginal orifice and adopted the squatting position. Contractions occurred at 2 to 3 minute intervals and each lasted 30 to 40 seconds. Labour lasted approximately one hour before the female punctured the water bag with her fingers and licked and handled the infant's head before complete delivery 3 to 4 minutes later. The newborn was immediately cradled and groomed. The placenta was delivered 25 minutes later and consumed. Female vervets in the field do not separate themselves from their troops before parturition (Gartlan 1969). The placenta is normally eaten (Gartlan 1969; Rowell 1970; Mallinson 1971).

Litter size

The litter size is usually 1 (Kushner *et al.* 1982) but twins are born very occasionally (Stott 1946; Hendrickx and Giles Nelson 1971).

Adult weight

Records of the adult weight of females range from 2.3 to 5.0 kg with most weighing 3.5 to 4.5 kg (Luck and Keeble 1967; Hopf and Claussen 1971; Bramblett and Coelho 1987; Happel *et al.* 1987). Males range from 3.1 to 7.0 kg with an average of about 4 to 5 kg (Poirier 1972). However, there is variation between strains (Luck and Keeble 1967), and on St. Kitts adult females and males averaged 8 kg and 11 kg respectively.

Neonate weight

The mean full-term birth weights of 24 males and 14 females were 328 g and 308 g respectively (Johnson *et al.* 1973). A birth weight below 250 g was considered dangerously low or a sign of prematurity (Hendrickx and Giles Nelson 1971).

Adult diet

In the wild, *C. aethiops* feed on leaves, fruit, flowers, herbs, shrubs, grass, seeds and crops. They also take animal matter, such as insects, nestlings, eggs, and rodents (Hall and Gartlan 1965; Poirier 1972; Lee *et al.* 1986). In one report on the feeding ecology of three groups of vervets in Senegal (Galat and Galat-Luong 1978), 34.4 per cent of the diet consisted of fruit of the following species of trees: *Acacia*, *Piliostigma*, *Ziziphus*, *Ficus*, and *Icacina*. Other plant material, such as shoots, leaves, grass, herbs, buds, flowers, fresh thorns, rhizomes, and gums, formed 38.7 per cent of the diet. The rest consisted of animal matter, such as grasshoppers, caterpillars, chrysalids, bird eggs, crabs, rats, hares, doves, hornbills, and sparrows.

In captivity, vervets will accept brown bread, a variety of fruits (apple, orange, pear, banana, strawberry, cherry, grapefruit) also cucumber, onion, tomato, potato (Mallinson 1970, 1971). They will also take corn, peanuts, and mangos but have been maintained on commercial monkey chow and *ad libitum* water (Johnson *et al.* 1980).

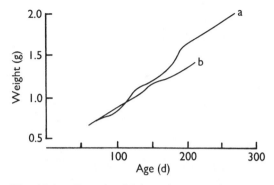

Fig. 11.1. Growth of (a) mother-reared *Cercopithecus pyargus* and (b) hand-reared *C. aethiops* vervet monkeys. From Mallinson (1971) and Tucker, (1985), respectively.

Kushner *et al.* (1982) fed a proprietary monkey chow with a daily supplement of apples. It has been reported that marrows and pumpkins cause diarrhoea, even when cooked (Luck and Keeble 1967).

Growth

There appears to be scant information on the growth rate of this species. The body weight of males increases from birth to maturity by a factor of 12, whereas that of females increases by a factor of 8 (Kretschmann *et al.* 1971). The growth curves of a mother-reared *Cercopitheus pyargus* and a hand-reared *C. aethiops* baby from 2 to 9 months are shown in Fig. 11.1 (Mallinson 1970; Tucker 1985).

Milk and milk intake

There are no data on the composition of the milk of *C. aethiops* as far as we are aware, but Jenness and Sloan (1970) provide some information on one sample taken from a green monkey (*C. sabeus*). The dry matter content was 16.4 per cent and the fat, protein, carbohydrate, and ash content of the fresh milk was 4.0 per cent, 3.1 per cent, 10.2 per cent and 0.6 per cent respectively. Therefore, the proportions of fat, protein, carbohydrate, and ash in the dry matter were 0.22, 0.17, 0.57, and 0.03 respectively. This is remarkably similar to the composition of human milk. Milk formulae designed for human infants are therefore probably very satisfactory, and both Enfamil and Similac

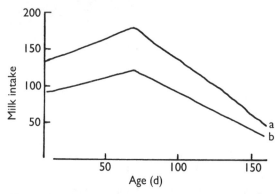

Fig. 11.2. Approximate daily (a) milk, ml/d, and (b) energy intake, kcal/d, provided by milk in hand-reared infant vervet monkeys. From Tucker (1985).

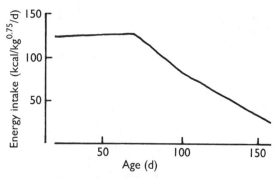

Fig. 11.3. Approximate daily energy intake provided by milk in hand-reared infant vervet monkeys per unit metabolic weight. From Tucker (1985).

with Iron have been used successfully (Tucker 1985; Maruska 1985).

Some information on milk consumption in relation to age in hand-reared infants is provided by Tucker (1985) and Maruska (1985). Milk intake appears to rise to a plateau at about 200 ml/d, which is consistent with an energy intake of about 120 kcal/d (Fig. 11.2). This represents a low energy intake rate (about 120 kcal/d per metabolic weight, compared with other species during growth (Fig. 11.3). Milk intake may fall from about 4 months of age as the amount of solid food ingested increases.

Lactation and weaning

Strong rooting and sucking behaviour has been observed within half an hour of birth (Shively and Mitchell 1986). Attachment to a nipple may act as a fifth anchoring point for the infant during its mother's movements (Shively and Mitchell 1986). In captivity, infants first take solids at about 2 months (Buss 1971), but four infants in the field were observed testing solids as early as 3, 9, 12, and 15 days of age respectively (Mallinson 1970, 1971). Mother-reared infants in captivity may be weaned at 6 months of age (Kushner et al. 1982).

Feeding

As far as we are aware there are no detailed accounts of the hand-rearing of neonatal vervet monkeys. Milk formulae designed for human infants are probably adequate, and the notes on feeding and management of young macaques are of relevance also to vervets.

There are brief notes on the hand-rearing of infant C. aethiops by Tucker (1985) and Maruska (1985) in the AAZPA Infant diet/care notebook. In one case, a day-old infant was fed four times daily and twice at night initially and this was reduced to twice daily and twice nightly at the age of 3 months. Cooked vegetables and fruit were gradually introduced after day 10. In the second report, a 2-month-old infant was reared using a baby bottle with a cross-cut nipple, and was fed six times daily, reduced to five times daily at 4 months, three times daily at 5 months, twice daily at 6 months, and once daily at 1 months. Monkey biscuit was offered by the age of 4 months and the infant was fully weaned by the age of 7.5 months. Abidec vitamin supplement was also added to the diet.

Accommodation

No specific data are available. Requirements are probably broadly similar to those described for macaques.

Infant management notes

There are no specific recommendations for this species. There is some evidence that neonatal mortality is lower in females that have previous breeding experience. Juvenile females show a great interest in neonates born in the troop and assist in their care. This behaviour is highly likely to be beneficial for subsequent care of their own babies, and colony management should allow young females to gain experience in this way.

The presence of young females acting as 'aunts' may also be directly beneficial to infants. Mallinson (1971) observed that infant *C. pyargus* born in groups that did not have juvenile female 'aunts', appeared to be very stressed by occasional neglect by their mothers.

Peer-group play (Fedigan 1972) and mixed-age grouping (Bloomstrand and Maple 1984) are probably very important for normal behavioural development. The principles outlined in the management of young rhesus macaques (Chapter 12) are probably equally relevant to this species.

Physical development

Cercopithecus aethiops infants are born with their eyes open (Rowell 1984; Shively and Mitchell 1986). The natal coat is very dark or black and the face is pink. This natal coat is replaced by the adult pelage at 2 months of age (Poirier 1972). The development of the infant vervet is relatively rapid (Rowell 1970), the hand and toe grip strengthens rapidly after birth and the infant becomes more robust after the first month (Johnson *et al.* 1980). The deciduous dentition consists of two incisors, one canine, and two molars on either side and in both upper and lower jaws. In the permanent dentition the two molars are replaced by two premolars and three molars. The eruption sequence for the permanent dentition is as follows: first molars, first incisors, second incisors, second molars, fourth premolars, third premolars, canines, and third molars, with the lower teeth usually erupting before the corresponding upper ones (Swindler 1976). Canine maturation was estimated to be complete in male vervets at 5 years of age (Bramblett and Coelho 1987).

Behavioural development

Newborn *C. aethiops* exhibit strong rooting reflexes and suck strongly within 30 minutes of birth (Shively and Mitchell 1986). The infant can cling soon after birth but the mother may help it to find the right position to suck (Gartlan 1969). It is carried on its mother's ventrum and may attach its mouth to both nipples for security (Mallinson 1971; Shively and Mitchell 1986). By 4 weeks of age infants still cling ventrally and dorsal carrying is rarely observed (Gartlan 1969). By 2 months of age the infants eat on their own and become much more mobile (Johnson *et al.* 1980). They spend the first three months of life close to their mothers who then stop carrying them except when danger threatens. By 6 to 7 weeks of age, infants in the wild spend most of their time with other juveniles of the same troop rather than their mothers (Lancaster 1971). The timing of infant independence varies from 4 to 8 months between troops, depending on habitat (McGuire *et al.* 1974)

The development of behaviour of juveniles within a social group was studied by Bramblett and Coelho (1987). The time spent in play decreases as puberty is approached. Grooming rates increase from birth to 87 months. Females exhibited adult grooming behaviour at 7 years. Males showed an increase in grooming rates for the first three years and behaved as adults thereafter. Aggression in females was highest when nearing their second birthday and then decreased to the adult level. Males reached their maximum aggression levels at 3 years of age.

Disease and mortality

There are no data on infant mortality rates in the field. In captive colonies abortion and still birth rates have been reported at 13 per cent (Hendrickx and Giles Nelson 1971), 16 per cent (35/221) (Kushner *et al.* 1982), 22 per cent (Fairbanks and McGuire 1984), and 28 per cent (Johnson *et al.* 1973) of pregnancies.

Of the 186 babies liveborn in the colony of Kushner *et al.* (1982), 8 (4 per cent) died within the first 3 days, 2 (1 per cent) between 3 and 10 days, 6 (3 per cent) between 10 and 50 days, and 7 (4 per cent) between 50 and 150 days. No deaths occured in juveniles older than 150 days. Significantly more males (17) died during the first 150 days than females (6) (Kushrer *et al.* 1982). Although the difference was not significant, these authors also noted a lower infant mortality rate at second births in captivity (10.5 per cent of 86), compared with first births (14 per cent of 100).

In another colony, the mortality rate for infants

under 1 month of age was 13 per cent, and that for infants between 1 and 13 months of age was 6 per cent (Fairbanks and McGuire 1984).

Preventative medicine

Johnson *et al.* (1973) examined pregnant females by rectal palpation at monthly intervals to monitor progress, and twice weekly in the fortnight before predicted parturition date to check the fetus was in the head-down position. Those in the breech position were manually rotated. The need for this high level of intervention was not made clear.

Indications for hand-rearing

There are no special indications for hand-rearing vervets.

Reintegration

There is no specific information on reintroducing hand-reared vervets to other groups.

References

Andelman, S., Else, J.G., Hearn, J.D., and Hodges, J.K. (1985). The non-invasive monitoring of reproductive events in wild vervet monkeys (*Cercopithecus aethiops*) using urinary pregnanediol-3-glucuronide and its correlation with behavioural observations. *Journal of Zoology, London (A)*, **205**, 467–77.

Asanov, S.S. (1972). Comparative features of the reproductive biology of Hamadryas baboons (*Papio hamadryas*), grivet monkeys (*Cercopithecus aethiops*) and rhesus monkeys (*Macaca mulatta*). In *The use of non-human primates in research on human reproduction* (ed. E Diczfalusy and C.C. Standley), pp. 458–72. WHO Research and Training Centre on Human Reproduction, Karolinska Institute, Stockholm.

Bloomstrand, M. and Maple, T.L. (1987). Management and husbandry of African monkeys in captivity. In *Comparative behaviour of African monkeys. Monographs in primatology*, Vol. 10 (ed. E.L. Zucker), pp. 197–235. Alan R. Liss, New York.

Bramblett, C.A. and Coelho, A.M. Jr. (1987). Development of social behaviour in vervet monkeys, Syke's monkeys and baboons. In *Comparative behaviour of African monkeys. Monographs in primatology*, Vol. 10 (ed. E.L. Zucker), pp. 67–79. Alan R. Liss, New York.

Bramblett, C.A., Pejaver, L.D., and Drickman, D.J. (1975). Reproduction in vervet and Syke's monkeys. *Journal of Mammology*, **56**, 940–6.

Buss, D.H. (1971). Mammary glands and lactation. In *Comparative reproduction of nonhuman primates* (ed. E.S.E. Hafez), pp. 315–33. Charles C. Thomas, Springfield, Illinois.

Fairbanks, L.A. and McGuire, M.T. (1984). Determinants of fecundity and reproductive success in captive vervet monkeys. *American Journal of Primatology*, **7**, 27–38.

Fedigan, L. (1972). Social and solitary play in a colony of vervet monkeys (*Cercopithecus aethiops*). *Primates*, **13**, 347–64.

Galat, G. and Galat-Luong, A.N.H. (1978). Diet of green monkeys in Senegal. In *Recent advances in primatology, Vol. 1. Behaviour* (ed. D.J. Chivers and J. Herbert), pp. 257–8. Academic Press, London.

Gartlan, J.S. (1969). Sexual and maternal behaviour of the vervet monkey, *Cercopithecus aethiops*. *Journal of Reproduction and Fertility*, Suppl. 6, 137–50.

Hafez, E.S.E. (1971). Reproductive cycles. In *Comparative reproduction of nonhuman primates* (ed. E.S.E. Hafez), pp. 85–114. Charles C. Thomas, Springfield, Illinois.

Hall, K.R.L. and Gartlan, J.S. (1965). Ecology and behaviour of the vervet monkey, *Cercopithecus aethiops*, Lolui Island, Lake Victoria. *Proceedings of the Zoological Society of London*, **145**, 37–56.

Happel, R.E., Noss, J.F., and Marsh, C.W. (1987). Distribution, abundance and endangerment of primates. In *Primate conservation in the tropical rain forest* (ed. C.W. Marsh and R.A. Mittermeier), pp. 63–82. Alan R. Liss, New York.

Hendrickx, A.G. and Giles Nelson, V. (1971). Reproductive failure. In *Comparaive reproduction of nonhuman primates* (ed. E.S.E. Hafez), pp. 403–25. Charles C. Thomas, Springfield, Illinois.

Henzi, S.P. and Lucas, J.W. (1980). Observations on the intertroop movement of adult vervet monkeys (*C. aethiops*). *Folia Primatologica*, **33**, 220–35.

Hopf. A. and Claussen, C.P. (1971). Comparative studies on the fresh weights of the brains and spinal cords of *Theropithecus gelada*, *Papio hamadryas* and *Cercopithecus aethiops*. *Proceedings of the Third International Primatology Congress*, Vol. 1, pp. 115–21. Zurich, 1970. Karger, Basel.

Horrocks, J.A. (1986). Life history characteristics of a wild population of vervets (*Cercopitrecus aethiops sabeus*) in Barbados. *International Journal of Primatology*, **7**, 31–47.

Jenness, R. and Sloan, R.E. (1970). The composition of milks of various species: a review. *Dairy Science Abstracts*, **32**, 599–612.

Johnsen, D.O. and Whitehair, L.A. (1986). Research facility breeding. In *Primates. The road to self-sustaining populations* (ed. K. Benirschke), pp. 499–511. Springer-Verlag, New York.

Johnson, C., Koerner, C., Estrin, M., Duoos, D. (1980). Allopaternal care and kinship in captive social groups of vervet monkeys (*Cercopithecus aethiops sabaeus*). *Primates*, **21**, 406–15.

Johnson, P.T., Valerio, D.A., and Thompson, G.E. (1973). Breeding African green monkeys *Cercopithecus aethiops* in a laboratory environment. *Laboratory Animal Science*, **23**, 355–59.

Jones, M.L. (1968). Longevity of primates in captivity. *International Zoo Yearbook*, **8**, 183–94.

Kretschmann, H.J., Schleifenbaum, L., and Wingert, F. (1971). Quantitative studies on the post-natal development of the central nervous system of *Cercopithecus aethiops*. *Proceedings of the Third International Primatology Congress, Vol. 1, pp. 108–14. Zurich, 1970. Karger, Basel.

Kushner, H., Kraft-Schveyer, N., Angelakos, E.T., and Wudarski, E.M. (1982). Analysis of reproductive data in a breeding colony of African green monkeys. *Journal of Medical Primatology*, **11**, 77–84.

Lancaster, J.B. (1971). Play-mothering: the relations between juvenile females and young infants among free-ranging vervet monkeys (*Cercopithecus aethiops*). *Folia Primatologica*, **15**, 161–82.

Lee, P.C., Brennan, E.J., J.G., and Altmann, J. (1986). Ecology and behaviour of vervet monkeys in a tourist lodge habitat. In *Primate ecology and conservation* (ed. J.G. Else and P.C. Lee), pp. 229–35. Cambridge University Press.

Luck, C.P. and Keeble, S.A. (1967). African monkeys. In *UFAW handbook on the care and management of laboratory animals* (3rd edn) pp. 734–42. Livingstone, Edinburgh and London.

Mallinson, J.J.C. (1970). Observations on the reproduction and development of the vervet monkey with special reference to intersubspecific hybridization. *Report of the Jersey Wildlife Preservation Trust*, **7**, 20–31.

Mallinson, J.J.C. (1971). Observations on the reproduction and development of vervet monkeys with special reference to intersubspecific hybridization. *Mammalia*, **35**, 598–609.

Maruska, E. (1985). Green guenon. In *Infant diet/care notebook*. (ed. S.H. Taylor, and D.H. Bietz). American Association of Zoo Parks and Aquariums, Wheeling, Virginia.

McGuire, M. (1974). Historical, ecological and population details. *The St Kitts vervet, contributions to primatology*. Vol. 1, pp. 5–27. Karger, Basel.

McGuire, M., et al. (1974). Specific behaviours. *The St Kitts vervet, contributions to primatology*, Vol. 1, pp. 92–129. Karger, Basel.

Michael, R.P. and Zumpe, D. (1971). Patterns of reproductive failure. In *Comparative reproduction of nonhuman primates* (ed. E.S.E. Hafez), pp. 205–42. Charles C. Thomas, Springfield, Illinois.

Mitchell, G. (1979a). Sexual behaviour: other Old World monkeys, apes, humans, generalisations and speculations. In *Behavioural sex differences in nonhuman primates*, pp. 140–64. Van Nostrand Reinhold, New York.

Mitchell, G. (1979b). Infant care: other Old World monkeys, apes and humans. In *Behavioural sex differences in nonhuman primates*, pp. 199–218. Van Nostrand Reinhold, New York.

Poirier, F.E. (1972). The St Kitts green monkey (*Cercopithecus aethiops sabaeus*): ecology, population dynamics, and selected behavioural traits. *Folia Primatologica*, **17**, 20–55.

Rowell, T.E. (1970). Reproductive cycles of two *Cercopithecus* monkeys. *Journal of Reproduction and Fertility*, **22**, 321–38.

Rowell, T.E. (1984). Guenons, macaques and baboons. In *The encyclopaedia of mammals*, Vol. 1 (ed. D. MacDonald), pp. 370–81. George, Allen & Unwin, London.

Rowell, T.E. & Richards, S.M. (1979) Reproductive strategies of some African monkeys. *Journal of Mammology*, **60**, 58–69.

Shively, C. and Mitchell, G. (1986). Perinatal behaviour of anthropoid primates. In *Comparative primate biology*, Vol. 2A. *Behaviour conservation and ecology* (ed. G. Mitchell and J. Erwin), pp. 245–94. Alan R. Liss, New York.

Stott, K. (1946). Twins in green guenon. *Journal of Mammology*, **27**, 394.

Swindler, D.R. (1976). *Dentition of living primates*, pp. 116–51. Academic Press, London.

Tucker, B.L. (1985). Grivet monkey. In *Infant diet/care notebook* (ed. S.H. Taylor and A.D. Bietz). American Association of Zoo Parks and Aquariums, Wheeling, Virginia.

Wolfheim, J.H. (1983). *Primates of the world. Distribution, abundance and conservation*, pp. 361–72. University of Washington Press, Seattle.

12 Rhesus macaque

Species

The rhesus macaque   *Macaca mulatta*

ISIS No.   1406008003006001

Status, subspecies, and distribution

The rhesus monkey has a larger geographical distribution than almost any other primate (Smith *et al.* 1987). Its range extends from India and Afghanistan in the west to China and Vietnam in the east (Roonwal and Mohnot 1977). It is not listed in the IUCN red data book (IUCN 1990). In India the population is reduced but still common in the north (Southwick and Lindburg 1986).

This species is widely used in biomedical research and Johnsen and Whitehair (1986) estimated that in 1984 the number of actual and potential breeding individuals at principal US research institutions was 24 934 and that the number of live births was 5582 (compared to 991 in 1973). The zoo population by comparison is small (Flesness 1986).

Sex ratio

At birth, sex ratios are thought to be close to 1:1 (Roonwal and Mohnot 1977). Results from captive colonies support this, for example, Hird *et al.* (1975) report 361 male to 376 female births, Rawlins and Kessler (1986) 583 male to 538 female births, and Bernstein and Gordon (1977) report 84 male to 78 female births. Bernstein and Gordon (1977) mentioned the mortality was greater among males than females during growth, especially about the time of puberty, but Hird *et al.* (1975) found no significant difference between sexes in post-natal mortality.

Social structure

In the wild, rhesus macaques live in multi-male troops whose size varies with habitat. Southwick *et al.* (1965) reported an average group size of 11 (range 1–50), Lindburg (1971) reported groups 8 to 98, and Seth and Seth (1986) observed up to 859 in and around ancient Indian temples.

Many rhesus macaques in captivity are housed in large groups in corrals. These are typically stocked at a density of about 40 animals per 2000 m^2 (Goodwin 1986). Paternity tests have shown that many males, including low-ranking adolescents, sire infants in a group (Bernstein and Gordon 1977). In captivity, smaller groups usually consist of one male and several females. Guidelines on forming new groups in captivity are given by Whitney and Wickings (1987).

Breeding age

The male shows active spermatogenesis at 3 years old and is sexually mature at 3–3.5 years old. However, breeding begins at 4.5–5 years of age and depends upon social maturity (Hafez 1971). Rhesus females show the first sexual skin swelling at about 26 months and first oestrus activity during the third breeding season after birth (Haddidian and Bernstein 1979). They can give birth at 3 to 4 years of age.

Longevity

The lifespan in captivity is typically 20–30 years. Jones (1980) cites a zoo record of 23 years and 8 months, and breeding can continue to 20 years of age or more (Hafez 1971). In a group at the Yerkes Primate Centre 3 (9.4 per cent) reached 31 years but the median lifespan was 15 years (Tigges *et al.* 1986).

Seasonality

The rhesus macaque is a seasonal breeder, with births occurring from March to June but mainly in April and May (Lindburg 1971). However, births can occur throughout the year in captivity (Hird *et al.* 1975). The duration of the oestrous cycle averages 28 days. Overt menstrual bleeding lasts 3 days and ovulation occurs 10–16 days after the first day of vaginal bleeding (Hendrickx and Giles Nelson 1971).

Interbirth intervals are close to one year on average, but tend to be shorter after a birth late in the year (June) than after an early birth (January), and are also shorter following abortions, still births, or the loss of a baby soon after birth (Goo and Fugate 1984).

Gestation

The gestation period averages 166 days (Whitney and Wickings 1987), and the range in 12 pregnancies reported by Michejda and Watson (1979) was 148 to 172 days.

Pregnancy diagnosis

Pregnancy can be diagnosed by palpation of the abdomen after 60 days of gestation. Radiography reveals the skeleton of the fetus in the later stages of pregnancy but, more appropriately, ultrasound scanning can be used. Pregnancy can be detected by ultrasonography at 16–18 days in rhesus macaques, and the fetal heart can be observed at 21–25 days (Tarantal and Hendrickx 1988). A test for detecting pregnancy from a small sample of urine (by assay for chorionic gonadotrophin), is available (Whitney and Wickings 1987).

Birth

Both in the wild and in captivity, birth occurs during the night or in the early morning (Lindburg 1971; Ruppenthal 1979). The female may vocalize and be restless during labour and usually adopts a squatting position (Bo 1971). The female eats the placenta and has usually cleaned the baby by daylight (Boelkins 1962). The duration of labour varies with parity and the presentation of the fetus. Multiparous females normally deliver their infants within 30 minutes, whereas primipara may take 1 to 3 hours. The mother assists in the delivery by pulling on the infant's head (Brandt and Mitchell 1971).

Imminent parturition may be assessed by rectal palpation to monitor fetal presentation and movements. Vaginoscopic examination will reveal the external os of the cervix dilating during the last 12–24 hours before delivery, so that the chorioamniotic membranes become visible (Mahony and Eisele 1978).

Litter size

The litter size is usually one. Twins are very rare. Only 4 out of 840 deliveries recorded by Hendrickx and Giles Nelson (1971) were twins.

Adult weight

Wild adult males weight between about 5.5 and 11 kg and wild adult females weigh between 3 and 10.5 kg (Roonwal and Mohnot 1977). The growth curves of Bourne (1975) show males reaching a mean weight of about 11 kg and females of about 8 kg at 7 years of age.

Neonate weight

The mean neonate weights of male and female rhesus macaques born at one colony were 513 g and 475 g respectively. The ranges amongst these 162 males and 151 females were 358–806 g and 287–678 g (Ruppenthal 1979). There appear to be significant differences in neonate weight between colonies. The means for 731 males and 724 females born at another centre were 488 g and 462 g respectively (Ruppenthal 1979), a few grams more than the means for males and females found by Van Wagenen (1972). Means recorded by Michejda and Watson (1979) for 4 males and 5 females were 451 g and 464 g and the mean weight of viable infants was 517 g (King and Chalifoux, 1986). Ruppenthal (1979) postulated that differences in gestation period between colonies resulting from environmental conditions might explain variation in neonate weight. Alternatively, differences in nutritional status may be involved. Another factor that can influence neonate weight is inbreeding. D.G. Smith (1986) found that inbred macaques had a significantly lower mean birth weight than non-inbred animals (441 g versus 497 g respectively).

Adult diet

In the wild, rhesus macaques are largely herbivorous, feeding on leaves, flowers, fruit, and seeds, although insects and other invertebrates are sometimes taken (Lindburg 1971). In captivity, proprietary primate pellets (with protein concentrations of 15 to 25 percent) usually form the basis of the diet and are fed *ad libitum*. These are supplemented with a variety of fruits and vegetables and, at some establishments, vitamin and mineral supplements (Ruppenthal 1979).

Adult energy requirements

The findings of several workers indicate that the adult rhesus macaque needs about 40–50 kcal gross energy per kilogram daily (National Research Council 1978). If we assume that the metabolizability (metabolizable energy density/gross energy density) of the diet is 0.75, and the weight of an adult is 8 kg, then this predicts a daily metabolizable energy requirement of 67 to 84 kcal/kg$^{0.75}$. This may be too low.

The results of Kemnitz (1986) indicated *ad libitum* intakes of about 100 to 233 kcal/kg$^{0.75}$ daily for animals weighing 10 kg.

Growth

Weight gain in captive rhesus macaques has been recorded by several workers (Van Wagenen and Catchpole 1956; Kirk 1972; Bourne 1975; Saxton and Lotz 1990). Their records indicate that in captivity the growth rate of female rhesus macaques gradually declines throughout growth and a weight of about 7.5 kg is typically attained by 7 years of age (Fig. 12.1). Males show a spurt in the rate of growth beginning at about 2.5 years of age and reach about 11 kg by 7 years of age (Fig. 12.1).

The growth of the fetus has been described by Van Wagenen (1972). Fetal weight gain is approximately linear from about 50 g at 80 days of gestation to about 450 g at 160 days of gestation: an average gain of 5 g/d.

Infants taken for artificial rearing may show a weight loss for two days before regaining birth weight at about 4 days old (Kerr *et al.* 1969b). However, the growth curves presented by Ruppenthal (1979) show that this is not always the case and that males, at least, tend to gain weight from the first day. There is wide individual variation. Kerr *et al.* (1969b) reared a group on

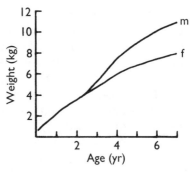

Fig. 12.1. Average growth curves of male (m) and female (f) rhesus macaques in captivity. From Bourne (1975).

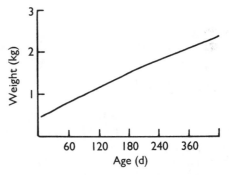

Fig. 12.2. Average growth of artificially reared rhesus macaques during the first year. From Kerr *et al.* (1969b).

Similac only for a year and those animals reached a mean weight of 2.35 kg by the end of the study (Fig. 12.2). The early growth of mother-reared infants may vary considerably and is influenced by birth weight, the mother's diet and lactation, environmental stresses, and the infant's health. Artificially reared infants often have higher growth rates (Michejda and Watson 1979).

Kerr *et al.* (1969a) showed that infants reared under conditions of total social isolation developed gross behavioural abnormalities, but grew at normal rates when fed *ad libitum*. Growth rates do not differ significantly between outbred (hybrid Chinese × Indian) strains and non-hybrid strains (Smith *et al.* 1987).

Milk and milk intake

There appears to be little information on the milk composition of rhesus macaques. Both Jenness and

Sloan (1970) and Ben Shaul (1962) cite the findings of Van Wagenen published in 1941. The dry matter content of 42 samples from 9 animals was found to be 12.2 per cent, and the milk contained 3.9 per cent fat, 2.1 per cent protein, 5.9 per cent carbohydrate, and 0.26 per cent ash. The proportions of fat, protein, carbohydrate, and ash in the dry matter were: 0.32, 0.17, 0.48, and 0.02 respectively. This composition is close to that of human milk and human milk replacers. For example, the composition of the dry matter of SMA is 0.28 fat, 0.12 protein, 0.55 carbohydrate, and 0.03 ash.

Human milk replacers have been found to be quite adequate for rearing infant rhesus macaques (although Ruppenthal 1979, reported two cases of lactose intolerance in babies fed human milk replacers). Kerr et al. (1969b) reared infants on Similac supplemented only with a drop of Paladec each day. In his review, Ruppenthal (1979) lists two institutions which reared infants on Similac and two which use SMA. One of those using Similac and one of those using SMA added 3–4 drops of a liquid multivitamin preparation each day until the babies were 90 days old. The milk replacers were not supplemented at the other two institutions. Unsupplemented human milk replacers seem to be adequate. It is of historic interest to note that rhesus infants have also been reared on a formula made from 1200 ml cow's milk, 600 ml water, 3.5 tablespoons of sucrose, and a vitamin supplement (Boelkins 1962).

The literature indicates that it has been common practice to offer a milk-free diet, such as 10 per cent dextrose solution, for the first few feeds or until day 2, and then to gradually introduce the milk replacer, so that by 2 to 4 days of age the infant is on full-strength milk replacer (Kerr et al. 1969b; Ruppenthal 1979). Other foods have been used to feed newborn babies, for example, premature infant formula or Dextri-Maltose(R) Ruppenthal (1979). Boelkins (1962) fed the babies he reared on a mixture of one part cow's colostrum to two parts of his diluted cow's milk formula (see above) for the first 5 days.

Milk intake at 2 days of age varies (Fig. 12.3) from 50 to over 150 ml, but by 14 days of age the mean intake of infants fed *ad libitum* is about 200–

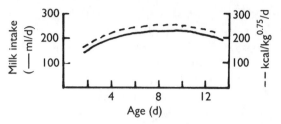

Fig. 12.3. Approximate mean daily milk intake (solid line) and energy intake provided by milk per metabolic weight (broken line) in hand-reared neonatal rhesus macaques. From Ruppenthal (1979).

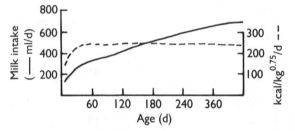

Fig. 12.4. Daily milk (solid line) and energy intake provided by milk per metabolic weight (broken line) in infant macaques reared on milk for 1 year. From Kerr et al. (1969b).

250 ml/d (Kerr et al. 1969b; Ruppenthal 1979). The mean milk intake of the infants reared by Kerr et al. (1969b) is shown in Fig. 12.4. The daily metabolizable energy intake *ad libitum* fed animals rises to a mean of about 250 kcal/kg$^{0.75}$ by one week of age and remains close to this for the first year of life (Fig. 12.4).

Lactation and weaning

Although infant rhesus macaques begin tasting solid foods at about 2 weeks of age the weaning process does not really begin until the third or fourth month of life, when the mother begins to actively prevent sucking. By 10 to 11 months lactation draws to an end and all infants are fully weaned before the birth of the next baby. Goo and Fugate (1984) weaned colony-born infants by separating them from their dams and putting them in groups of 10–14 same-age peers. They found no significant differences in survival to 2 years among groups weaned at 6, 8, 10, and 12 months of age, although 1-year-old infants weaned at 12 months were 200 g heavier on average than those weaned at 6 months.

Techniques for weaning artificially reared infants vary between institutions, but those reviewed by Ruppenthal (1979) started providing monkey chow and fruit between 16 and 50 days after birth, and the age at which weaning was completed varied from 180 to 365 days. Lang (1971) found that infants took to solid foods readily and soon preferred them to milk formula. Water should, of course, be available *ad libitum*. The amount of milk provided should be reduced gradually. At one centre the milk concentration was reduced, after an infant had reached 90 to 135 days of age, by adding increasing volumes of water (Ruppenthal 1979).

Feeding

Infants have been fed using a 50 ml bottle manufactured for use in pet animals. A suitable sized hole in the nipple was made by piercing it with a red-hot 22 gauge needle (Ruppenthal 1979). The infant should be held in an upright position for feeding and wrapped in or placed on a diaper which it can grasp. Most can be trained to feed themselves from a bottle attached to the incubator or cage from 7 to 14 days of age (Kerr *et al.* 1969b; Ruppenthal 1979), or even by 3 to 4 days of age (Anderson 1986).

Kerr *et al.* (1969b) began feeding their animals 8 to 16 hours after birth, and fed at 2 hour intervals until 3 days of age. The frequency was then gradually reduced to 4 hour intervals by 12 days of age. The animals were fed *ad libitum*. Feeding regimes employed by others are generally similar to this but some limit food intake at each feed to a maximum of 25 ml or less (Ruppenthal 1979). Boelkins (1962) discontinued night feeds by 8 to 11 days after birth and reduced the number of feeds to 4 per day by one month, 3 per day by three months, and twice daily by 6 months. The animals can be weaned by 6 months, although there may be some advantage in continuing to provide some milk until a year old (Goo and Fugate 1984; Ruppenthal 1979).

Infants born more than 16 days before full term or weighing less than about 375 g were considered by Ruppenthal (1979) to be at extreme risk from aspiration pneumonia due to immaturity in their ability to co-ordinate sucking. Those were fed for up to 10 days via a 3.5-5.0 French infant feeding tube inserted orally or nasally into the stomach with small feeds, initially 3-5 ml, every 2 hours.

Accommodation

Newborn macaques separated from their mothers have been housed in Isolette incubators or other cages that enable the maintenance of a stable rectal temperature of 37 °C. Kerr *et al.* (1969b) kept the babies in individual cages in an environment of 22-23 °C and 35 per cent humidity, and provided heating pads for the first 15 days. This system worked, but humidity should preferably be kept above 50 per cent. Those that have used Isolettes have set them to provide temperature of 29-34 °C initially and have maintained a humidity of 60-90 per cent (Ruppenthal 1979). This author indicated that the temperature is generally reduced gradually until the animals are placed, at about 14 days old, in individual cages in a room temperature of 23-28 °C. He also found that it was common practice to line the floor of the rearing cage with diapers for warmth and to help prevent urine-scalding, although there was great variation between institutions in the age at which the diapers were no longer used.

Anderson (1986) reported that nursery-reared macaques at the California Regional Primate Center develop into successful mothers. Their babies are always paired or put into small groups by 3-4 weeks of age, and are moved into larger peer groups at 3-6 months of age, to provide socialization.

Infant management notes

The influence of rearing conditions on subsequent breeding and mothering ability has been studied extensively in the rhesus macaque, and it is well established that if weaned at birth and deprived of social interaction for 6 months or more months, individuals are very rarely capable of normal reproduction (Goldfoot 1977). Such animals typically show severe behavioural abnormalities including hyper-aggression, self-biting, self-clasping, rocking, stereotypical pacing, and catatonic-like states (Harlow et al. 1966; Goldfoot 1977).

Goldfoot (1977) suggested that 'for eventual reproductive potential to develop, rhesus infants needed complex social environments in which aggression and fear among peers occurs only at low levels and that the closer one came to duplicating the naturalistic social environment of the monkey, the better the chance one had of producing reproductively capable animals'.

Suomi (1986) described principles of management aimed at improving reproductive and maternal competence in a colony. We have attempted to summarized these briefly below.

1. Whenever possible, infants were reared by their own mothers, at least until weaning.
2. Babies neglected by their own mothers were fostered on to other females whenever possible.
3. Those babies that had to be taken for artificial rearing were placed in peer groups at as early an age as possible, and for a minimum of at least an hour several times a week during the first year of life.
4. When possible animals were maintained, from the juvenile stage through to adulthood, in stable social groups which contained peers and some related older animals which were reproductively active.
5. Females were not removed from the group to give birth.

Suomi (1986) reported that, at that time, over 90 per cent of primiparous nursery-reared females were successful mothers. In the earlier days of the colony, before the measures outlined above had been introduced, the chances of a nursery-reared primiparous mother rejecting her infant were about 80 per cent.

Physical development

The testes are in the scrotum at birth but reascend into the inguinal canal shortly afterwards (Kinzey 1971). They return permanently to the scrotum at about 3.3 years of age (Zimmerman et al. 1975).

In the animals studied by Michejda and Watson (1979), the eruption of the deciduous lower and upper central incisors was completed in the fourth week, the deciduous upper and lower lateral incisors erupted at 8 to 10 weeks of age, and the deciduous first molars were fully erupted 12 weeks after birth. These authors pointed out that these eruptions all took place about 2 weeks sooner than expected from earlier studies on the rhesus macaque and postulated that nutritional conditions may influence tooth eruption dates to some extent. The sequence of eruption of the permanent dentition is: lower first molars, upper first molars, upper and lower first incisors, upper second incisors, lower second molars, upper second molars, lower third premolars, upper third premolars, lower fourth premolars, upper fourth premolars, upper and lower canines, upper third molars, and finally lower third molars (Swindler 1976).

Because of the wide variation in growth rates between individuals associated with environmental conditions, measurement of linear dimensions or body weight can give only a rough and unreliable indication of age of this species. Accurate ageing by examination of the dentition is also difficult (Michejda and Watson 1979). These authors considered that the ossification of the hand and wrist can give the most accurate estimation of age.

Behavioural development

In the very early post-natal period the newborn rhesus monkey spends most of the time sleeping and sucking (Lindburg 1971). By 2 days of age, however, the infant can follow nearby movements with its eyes and can sit and crawl on the ground (Roonwal and Mohnot 1977). Initially, babies are carried in the ventral position but change to the dorsal position at 1–4 weeks of age (Lindburg 1971). By day 8 or 9 the infant begins to break contact with its mother, and first begins to show play behaviour at 16 days. By the third week it leaves its mother regularly on short excursions and by the sixth week the infant can travel on its own, although it may still be carried occasionally until about 5 months old (Lindburg 1971). The mother may restrict the movements of the baby until it is about 7 weeks old, but after that it plays freely with other monkeys (Roonwal and Mohnot 1977).

The impact of various husbandry procedures on behavioural development has been reviewed by Ruppenthal and Sackett (1979). Normal behavioural development is fragile and easily disrupted in captive conditions, and behavioural abnormalities are difficult to correct.

Disease and mortality

The mortality in wild rhesus macaques prior to two years of age was found to vary from 15 to 22 per cent in three populations in India and Nepal (Teas et al. 1980). Koford (1965) estimated that about 4 per cent of births in the free-living population at Cayo Santiago Island were stillborn, that mortality to 12 months of age was 6 to 11 per cent, and that for animals in their second year was 10 per cent. Analysis of the demographics of this population between 1976 and 1983 revealed that 46 (4 per cent) of 1157 births to females aged 4 years or older were abortions or still births, and that the average mortality during the first year was 7.3 per cent (Rawlins and Kessler 1986).

Goo and Fugate (1984) reported rather similar rates of mortality in colony-born animals: an overall mortality of 14 to 16 per cent up to 2 years of age, with mortality during the first 6 months at 8.3 per cent. These authors mentioned that prior to 2 years of age 30 to 35 per cent of their animals required clinical care, and that the main problems were enteropathies and injuries. The outcome of 401 pregnancies in one captive colony was 33 (8.5 per cent) abortion, 42 (10.5 per cent) stillborn, and only 15 (5 per cent of live births) died within a week of birth (Van Wagenen 1972).

Of 742 live births reported by Hird et al. (1975), 80 (10.8 per cent) died within 30 days of birth, and a total of 126 (17 per cent) in the first 6 months. In this study, neonatal mortality (from birth to 30 days) was found to be much higher in outdoor-born babies than in those born indoors (25 per cent versus 8 per cent), and there was also a marked difference in mortality rate between outdoor and indoor babies aged 1 to 6 months (16 per cent versus 6 per cent). Amongst those born in the outdoor accommodation there was a clear seasonal variation in neonatal mortality, it being highest in the winter months (38 per cent in February versus 5 per cent in September). Neonatal mortality (to 30 days) in artificially reared infants at a research center was 6.7 per cent (Johnson et al. 1986).

The social rank of the mother has been found to have a significant effect on infant survival in captivity (Wilson et al. 1978). Mortality to 6 months of age was 22.6 per cent in babies born to low-ranking mothers compared with 10.3 per cent in those born to higher ranks.

Padovan and Cantrell (1983) surveyed the necropsy reports of 144 mainly mother-reared rhesus infants that died within 18 months of birth: 50 (5 per cent) of 1045 liveborn died within 30 days of birth and a total of 144 (14 per cent) of the 1045 died before 18 months of age. Bronchopneumonia associated with a variety of bacterial agents was the apparent cause of death in 13 per cent of these animals and enteritis and/or colitis account for another 25 per cent. Trauma was not a common cause of death in the infants in this survey.

Low birth weight reduces the chances of survival, and Trum (1972) considered that those less than 350 g were at high risk. However, Ruppenthal (1979) showed that mortality amongst infants in the lightest 10 per cent of the weight range at birth dropped from 62 per cent in 1974 to 14 per cent in 1976 as a result of greater veterinary intervention and improvements in nursery management.

The mortality rates of artificially reared infants can be very much lower than those seen in wild or in colony-managed animals, and it is for this reason that artifical rearing is a quite widespread management system in laboratory-housed macaque colonies. Scheffler and Kerr (1975) reported rearing 50 infants, separated from their mothers within hours of birth, with no mortality during the first year of life. However, in one earlier large-scale artificial rearing programme 114 (19 per cent) of 593 infants died befoe 12 months of age (Valerio et al. 1969). King and Chalifoux (1986) found that those infants that survived the first 12 hour period between birth and transfer to the nursery had a less than 5 per cent mortality to 30 days of age and beyond.

Fetal losses in captive rhesus macaques have

been reported at 8.5 per cent and 9.1 per cent by Van Wagenen (1972) and Bernstein and Gordon (1977) respectively. Higher rates of up to 15 to 21 per cent were reported by Ruppenthal (1979), but this was reduced to 5 per cent with veterinary intervention. At one centre 181 (17 per cent) of 1076 births were stillborn (Johnson et al. 1986). In a survey of 21 late-term (more than 139 days of gestation) fetal deaths, Mahony and Eisele (1978) found that seven were breech births, 1 a frontum dystocia, 7 had placental abnormalities, and the cause of death of the other 6 was not established. Earlier fetal deaths and abortions were due mainly to maternal infections (e.g. Shigellosis). Mahony and Eisele (1978) described a pre-partum care programme which included regular examination of pregnant females after 150 days of pregnancy by rectal palpation and using ultrasonography, so that fetal malpresentations could be detected and corrected.

Preventative medicine

Regular testing and screening for tuberculosis has been a part of routine procedures in laboratory colonies. The skin test is performed by intradermal injection of 0.1 ml mammalian tuberculin (KOT) or purified protein derivative (PPD) containing 10–1000 Tuberculin Units into the upper eyelid with a 25 gauge needle. The site is examined 24, 48, and 72 hours after inoculation (Klos and Lang 1982; Vickers 1986). Some advocate active immunization with BCG vaccine (Klos and Lang 1982).

Rhesus macaques are susceptible to measles, and vaccination against this and poliomyelitis is recommended in primate colonies by Hunt (1986). However, there seems to be little concensus about vaccination protocols.

Exclusion of humans carrying potentially transmissible diseases is, of course, a fundamental aspect of disease prevention.

Indications for hand-rearing

Bernstein and Gordon (1977) expressed the view that although some weak infants can be salvaged from a colony by intervention and hand-rearing, in view of the fragility of sexual behaviour in artificial-rearing conditions, providing the environment needed for normal behavioural development is 'ordinarily beyond the capability of any laboratory not specifically designed for infant rearing studies'. In their experience, in the event of the death of a mother, grandmothers, maternal aunts or sisters are often able and willing to care for weaned infants, and pre-weaning babies have sometimes been successfully adopted by a lactating female.

If a baby is completely neglected by its mother then it should be collected as soon as possible and if it cannot be fostered to another female, it should be hand-reared or humanely destroyed. It is relatively easy to meet the physiological demands of an infant macaque and to rear it, but the task should not be undertaken without consideration of the animal's future. Unless it can be kept with others for socialization there is little merit in starting to hand-rear.

Reintegration

S. Smith (1986) described the results of an experiment to study the success of cross-fostering and reported an 80 per cent success rate for cross-fostering neglected or abused infants. Females with infants of up to 14 days of age were isolated from their social groups in small cages. Their own babies were removed and, soon afterwards, they were restrained whilst babies for fostering were placed on them. The mothers with their fostered infants were then observed until sucking occurred and were then released back into their social groups in the cornal. The fostered infants were accepted in 27 out of 36 cases. Hansen (1966) found that it may be three weeks before a mother is able to discriminate between her own and other babies. These results indicate that cross-fostering healthy abandoned or orphaned babies to other lactating females that have recently lost their own babies, stands a good chance of success in the neonatal period. S. Smith (1986) pointed out that this technique was valuable in introducing new animals in order to prevent inbreeding in established groups, in view of the difficulties of introducing older animals.

If no foster mother is available, hand-reared babies should be kept in small peer groups by 3 weeks of age and moved to larger peer groups at 3 to 6 months to provide socialization. Introduction of individuals to established groups is difficult.

References

Anderson, J.H. (1986). Rearing and intensive care of neonatal and infant nonhuman primates. In *Primates. The road to self sustaining Populations* (K. Benirschke), pp. 747–62. Springer-Verlag, New York.

Ben Shaul, D.M. (1962) The composition of the milk of wild animals. *International Zoo Yearbook*, **4**, 333–42.

Bernstein, I.S. and Gordon, T.P. (1977). Behavioural research in breeding colonies of Old World monkeys. *Laboratory Animal Science*, **27**, 532–40.

Bo, W.J. (1971). Parturition. In *Comparative reproduction of nonhuman primates* (ed. E.S.E. Hafez), pp. 302–13. Charles C. Thomas, Springfield, Illinois.

Boelkins, R.C. (1962). Large scale rearing of infant Rhesus macques (*Macaca mulatta*) in the laboratory. *International Zoo Yearbook*, **4**, 285–9.

Bourne, G.H. (ed.) (1975). *The rhesus monkey*, Vol. 1. Academic Press, New York.

Brandt, E.M. and Mitchell, G. (1971). Parturition in primates. In *Primate behaviour: developments in field and laboratory research*, Vol. 2 (ed. L.A. Rosenblum), pp. 117–23. Academic Press, New York.

Flesness, N. (1986). Captive status and genetic considerations. In *Primates. The road to self-sustaining populations* (ed. K. Benirschke), pp. 845–56. Springer-Verlag, New York.

Goldfoot, D.A. (1977). Rearing conditions which support or inhibit later sexual potential of laboratory-born rhesus monkeys: hypotheses and diagnostic behaviours. *Laboratory Animal Science*, **27**, 548–56.

Goo, G.P. and Fugate, J.K. (1984). Effects of weaning age on maternal reproduction and offspring health in rhesus monkeys (*Macaca mulatta*). *Laboratory Animal Science*, **34**, 66–9.

Goodwin, W.J. (1986). Corral breeding of nonhuman primates. In *Primates. The road to self-sustaining populations* (ed. K. Benirschke), pp. 289–95. Springer-Verlag, New York.

Haddidian, J. and Bernstein, I.S. (1979). Female reproductive cycles and birth data from an old world monkey colony. *Primates*, **20**, 429–42.

Hafez, E.S.E. (1971). Reproductive cycles. In *Comparative reproduction of nonhuman primates* (ed. E.S.E. Hafez), pp. 85–114. Charles C. Thomas, Springfield, Illinois.

Hansen, E.W. (1966). The development of maternal and infant behaviour in the rhesus monkey. *Behaviour*, **27**, 107–49.

Harlow, H.F., Harlow, M.K., Dodsworth, R.O., and Arling, G.L. (1966). Maternal behaviour of rhesus monkeys deprived of mothering and peer associations in infancy. *Proceedings of the American Philosophical Society*, **110**, 58–66.

Hendrickx, A.G. and Giles Nelson, V. (1971). Reproductive failure. In *Comparative reproduction of non-human primates* (ed. E.S.E. Hafez), pp. 403–25. Charles C. Thomas, Springfield, Illinois.

Hird, D.W., Henrickson, R.V., and Hendrickx, A.G. (1975). Infant mortality in *Macaca mulatta*: neonatal and post-natal mortality at the California Primate Research Center. *Journal of Medical Primatology*, **4**, 8–22.

Hunt, R.D. (1986). Viral diseases of neonatal and infant nonhuman primates. In *Primates. The road to self-sustaining populations* (ed. K. Benirschke), pp. 725–42. Springer-Verlag, New York.

IUCN (1990). *1990 IUCN red list of threatened animals*. IUCN, Gland, Switzerland.

Jenness, R. and Sloan, R.E. (1970). The composition of milks of various species: a review. *Dairy Science Abstracts*, **32**, 599–612.

Johnsen, D.O. and Whitehair, L.A. (1986). Research facility breeding. In *Primates. The road to self-sustaining populations* (ed. K. Benirschke), pp. 499–510. Springer-Verlag, New York.

Johnson, L.D., Petto, A.J., Boy, D.S., Sehgal, P.K., and Beland, M.E. (1986). The effect of perinatal and juvenile mortality on colony-born production at the New England Regional Primate Research Center. In *Primates: The road to self-sustaining populations* (ed. K. Benirschke), pp. 771–9. Springer-Verlag, New York.

Jones, M.L. (1980). Lifespan in mammals. In *The comparative pathology of zoo animals* (ed. R.J. Montali, and G. Migaki), pp. 495–509. Smithsonian Institution Press, Washington D.C.

Kemnitz, J.W. (1986). Energy balance in laboratory-housed rhesus monkeys. *Primate Report*, **14**, 149–50.

Kerr, G.R., Chamove, A.S., and Harlow, H.F. (1969a). Environmental deprivation: its effect on the growth of infant monkeys. *Journal of Pediatrics*, **75**, 833–7.

Kerr, G.R., Scheffler, G., and Waisman, H.A. (1969b). Growth and development of infant *Macaca mulatta* fed a standardised diet. *Growth*, **33**, 185–99.

King, N.W. Jr. and Chalifoux, L.V. (1986). Prenatal and neonatal pathology of captive nonhuman primates. In *Primates. The road to self-sustaining populations* (ed. K. Benirschke), pp. 763–70 Springer-Verlag, New York.

Kinzey, W.G. (1971). Male reproductive system and spermatogenesis. In *Comparative reproduction in nonhuman primates* (ed. E.S.E. Hafez), pp. 85–114. Charles C. Thomas, Springfield, Illinois.

Kirk, J.H. (1972). Growth of maturing *Macaca mulatta*. *Laboratory Animal Science*, **22**, 573–5.

Klos, H. and Lang, E. (1981). *Handbook of zoo medicine*. Reinhold, New York.

Koford, C.B. (1965). Population dynamics of rhesus monkeys on Cayo Santiago. In *Primate behaviour* (ed. I. Devore), pp. 160–74. Holt, Reinhart & Winston, New York.

Lang, M.C. (1971). Techniques of breeding and rearing of monkeys. In *Comparative reproduction of nonhuman primates* (ed E.S.E. Hafez), pp. 455–72. Charles C. Thomas, Springfield, Illinois.

Lindburg, D.G. (1971). The rhesus monkey in North India: an ecological and behavioural study. In *Primate Behavior: developments in field and laboratory research*, Vol 2 (ed. L.A. Rosenblum), pp. 1–106. Academic Press, New York.

Mahony, C.J. and Eisele, S. (1978). A programme of pre-partum care for rhesus monkey *Macacca mulatta*: results of the first two years of study. In *Recent advances in primatology*, Vol. 2. *Conservation*, (ed. D.J. Chivers and W. Lane-Petter), pp. 265–7. Academic Press, London.

Michejda, M. and Watson, W.J. (1979). Age determinants in neonatal primates: a comparison of growth factors. In *Nursery care of non-human primates* (ed. G.C. Ruppenthal), pp. 61–75. Plenum Press, New York.

National Research Council (1978). *Nutrient requirements of non-human primates*, pp. 34–6. Nutrient requirements of domestic animals, No. 14. National Academy of Sciences, Washington DC.

Padovan, D. and Cantrell, C. (1983). Causes of death in infant rhesus and squirrel monkeys. *Journal of American Veterinary Medicine Association*, **183**, 1182–4.

Rawlins, R.G. and Kessler, M.J. (1986). Demography of the free-ranging Cayo Santiago macaques (1976–1983). In *The Cayo Santiago macaques* (ed. R.G. Rawlins and M.J. Kessler), pp. 47–72. State University of New York Press, Albany.

Roonwal, M.L. and Mohnot, S.M. (1977). *Primates of South-East Asia*, pp. 97–171. Harvard University Press, Cambridge, Mass, and London.

Ruppenthal, G.C. (1979). Survey of protocols for nursery rearing infant macaques. In *Nursery care of nonhuman primates* (ed. G.C. Ruppenthal), pp. 165–85. Plenum Press, New York.

Ruppenthal, G.C. and Sackett, G.P. (1979). Experimental and husbandry procedures: their impact on development. In *Nursery care of nonhuman primates* (ed. G.C. Ruppenthal), pp. 269–84. Plenum Press, New York.

Saxton, J.L. and Lotz, G. (1990). Growth of rhesus monkeys during the first 54 months of life. *Journal of Medical Primatology*, **19**, 119–36.

Scheffler, G. and Kerr, G.R. (1975). Growth and development of infant *M. arctoides* fed a standardized diet. *Journal of Medical Primatology*, **4**, 32–44.

Seth, P.K. and Seth, S. (1986). Ecology and behaviour of rhesus monkeys in India. In *Primate ecology and conservation*, Vol. 2 (ed. J.G. Else and P.C. Lee), pp. 89–103. Cambridge University Press.

Smith D.G. (1986). Incidence and consequences of inbreeding in three captive groups of rhesus macaques (*Macaca mulatta*). In *Primates. The road to self-sustaining populations* (ed. K. Benirschke), pp. 857–74. Springer-Verlag, New York.

Smith, D.G., Lorey, F.W., Suzuki, J., and Abe, M. (1987). Effect of outbreeding on weight and growth rate of captive infant rhesus macaques. *Zoo Biology*, **6**, 201–12.

Smith, S. (1986). Infant cross-fostering in captive rhesus monkeys (*Macaca mulatta*). *American Journal of Primatology*, **11**, 229–37.

Southwick, C.H. and Lindburg, D.G. (1986). The primates of India: status trends, and conservation. In *Primates. The road to self-sustaining populations* (ed. K. Benirschke), pp. 171–87. Springer-Verlag, New York.

Southwick, C.H., Beg, M.A., and Siddigi, M.R. (1965). Rhesus monkeys in North India. In *Primate behaviour: field studies of monkeys and apes* (ed. I. DeVore), pp. 111–59. Holt, Rhinehart & Winston, New York.

Suomi, S.J. (1986). Behavioural aspects of successful reproduction in primates. In *Primates. The road to self-sustaining populations* (ed. K. Benirschke), pp. 331–40. Springer-Verlag, New York.

Swindler, D.R. (1976). *Dentition of the living primates*, pp. 129–32. Academic Press, New York, London.

Tarantal, A.F. and Hendrickx, A.G. (1988). Use of ultrasound for early pregnancy detection in the rhesus and cynomolgus macaque (*Macaca mulatta* and *Macaca fascicularis*). *Journal of Medical Primatology*, **17**, 105–12.

Teas, J., Richie, T., Taylor, H., and Southwick, C. (1980). Population patterns and behavioural ecology of the rhesus monkey (*Macaca mulatta*) in Nepal. In *The Macaques: studies in ecology, behaviour and evolution* (ed. D.G. Lindburg), pp. 247–62. Van Nostrand Reinhold, New York.

Tigges, J., Gordon, T.P., and McClure, H.M. (1986). Life span of rhesus monkeys (*Macaca mulatta*) in the Yerkes colony. *Primate Report*, **14**, 129.

Trum, B.F. (1972). Research on reproduction at the New England Regional Primate Research Center. In *Breeding primates* (ed. W.I.B. Beveridge), pp. 198–202. Karger, Basel.

Valerio, D.A., Miller, R.L., Innes, J.R.M., Courtney, K.D., Pallotta, A.J., and Guttmacher R.M. (1969). *Macaca mulatta: management of a laboratory breeding colony*. Academic Press, New York.

Van Wagenen, G. (1972). Vital statistics from a breeding colony: reproduction and pregnancy outcome in *Macaca mulatta*. *Journal of Medical Primatology*, **1**, 3–28.

Van Wagenen, G. and Catchpole, H.R. (1956). Physical growth of the rhesus monkey (*Macaca mulatta*). *American Journal of Physical Anthropology*, **14**, 245–73.

Vickers, J.H. (1986). Approaches to determining colony infections and improving colony health. In *Primates: the road to self-sustaining populations* (ed. K. Benirschke), pp. 521–30. Springer-Verlag, New York.

Whitney, R.A. and Wickings, E.J. (1987). Macaques and other Old World simians. In *The UFAW handbook on care and management of laboratory animals* (ed. T.B. Poole), pp. 599–627. Longman, Harlow.

Wilson, M.E., Gordon, T.P., and Bernstein, I.S. (1978). Timing of reproductive success in rhesus monkey social groups. *Journal of Medical Primatology*, **7**, 202–12.

Zimmerman, R.R., Strobel, D.A., Steeve, P., and Geist, C.R. (1975). Behaviour and malnutrition in the rhesus monkey. In *Primate behaviour: development in field and laboratory research*, Vol. 4 (ed. L.A. Rosenblum), pp. 241–306. Academic Press, New York.

13 Stump-tailed macaque

Species

The stump-tailed macaque   *Macaca arctoides*

ISIS No.   1406008003001002

Status, subspecies, and distribution

The range of *M. arctoides* extends from eastern Bangladesh eastwards through southern China and into the Malayan peninsula and Vietnam. It is declining, or has become rare in India, Thailand, and Malaysia but its status elsewhere in its range is unknown (Wolfheim 1983). It is not classified as threatened by the IUCN (1990).

The stump-tailed macaque is quite widely kept as a laboratory animal. Johnsen and Whitehair (1986) estimated the population of potential breeders in US research facilities to be 386 in 1984.

Sex ratio

Harvey and Rhine (1983) report a sex ratio of 40 males to 24 females in births at their colony, and cite Brüggerman and Grauwiler's (1972) report of a ratio of 39 males to 17 females. It appears that there may be a biased sex ratio at birth in this species in captivity.

Social structure

Wild stump-tailed macaques live in groups that vary in size according to habitat but are typically of 25 to 30 individuals (McCann 1933).

Breeding age

Females reach sexual maturity at approximately 3 years of age and males at 3.5 years of age (Trollope and Jones 1972). Captive animals appear to mature earlier than those in the wild.

Longevity

We are not aware of any specific data. Macaques appear to have maximum lifespans of about 25 to 30 years.

Seasonality

Female *M. arctoides* have oestrous cycles throughout the year and do not show a seasonal anovulatory period like that of *M. mulatta* (Stenger 1972; Dukelow 1973; Trollope 1978). Births, at least in captivity, are evenly distributed throughout the year (A. Estrada and B. Estrada 1976; Chamove 1981; Harvey and Rhine 1983). The duration of the cycle is 28 to 34 days (Stenger 1972) with a mean of 29.4 days (Trollope, 1978). Menstrual flow lasts 3.7 days but gross vaginal bleeding is seldom observed (Stenger 1972). There is no swelling of the perineal skin during oestrus (Michael and Zumpe 1971), but the animals' faces turn bright red (Chamove 1981).

The interbirth interval amongst mothers who lost their infants within 32 days of birth was found to average 10 months, whilst that for mothers who kept their infants for at least 8 months ranged from about 1 to 2.5 years, with a mean of about 1.5 years (Harvey and Rhine 1983). These authors also noted that interbirth intervals tended to increase with decreasing social rank and senility.

Gestation

The gestation periods of eight pregnancies for which conception date was known from observation of menstrual cycles and copulation, ranged from 169 to 195 days, with a mean of 180 days (Harvey and Rhine 1983). These figures agree closely with other reports (Stenger 1972; Trum 1972; Trollope 1978).

Pregnancy diagnosis

Most females show obvious abdominal swelling during pregnancy (Chamove 1981) and the sexual skin becomes a darker red than during menstrual cycling (Trollope and Jones 1975). Cessation of menstruation may indicate pregnancy. Bunyak *et al.* (1982) have described techniques of vaginal swabbing in gang-housed animals.

Birth

Births occur at night (Trollope 1978; Chamove 1981) and the placenta is probably eaten after parturition (Brandt and Mitchell 1971; Trollope 1978). Other members of the group show an interest in birth but do not intervene (Trollope 1978). Post-partum bleeding has been observed for up to 7 days (Trollope 1978), but should be investigated if heavy.

Litter size

The usual litter size is 1 but twins occur occasionally (Brandt and Mitchell 1971; Hendrickx and Giles Nelson 1971).

Adult weight

Adult weight varies considerably and is related to the environmental conditions in which the animals have been raised. Wild-caught adult males or females weighed 7 kg and 5 kg respectively, whereas those reared in captivity averaged 13.2 kg and 20 kg (Chamove 1981). Wide variation was also reported by Faucheux *et al.* (1978) from a mean of 7.6 kg in wild females to 12.4 kg in some reared in a zoo, and there were marked differences in linear dimensions as well as weight. Harvey *et al.* (1979) recorded mean weights of 15.2 kg for 8 males and 10.0 kg for females.

Neonate weight

From the graph (Fig. 13.1) presented by Chamove (1981) it appears that neonate weight averages close to 500 g and ranges from 400 to 600 g. The data of Scheffler and Kerr (1975) indicate a slightly lower average, and these authors found no significant difference in birth weight between sexes. The weight of females measured by Faucheux *et al.* (1978) was listed as 450 g at 1 week of age. Harvey *et al.* (1979) found the mean weights of 6- to 10-day-old babies were 480 g and 450 g for 16 males and 10 females respectively.

Adult diet and energy requirements

No specific data are provided (see species macaque, Chapter 12).

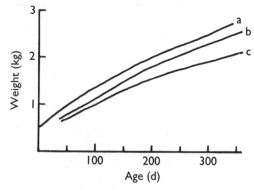

Fig. 13.1. Growth of infant stump-tailed macaques. (a) mean of both sexes. From Chamove (1981). (b) mean for males and (c) mean for females. From Scheffler and Kerr (1975).

Growth

Males, although almost the same weight as females at birth, grow more rapidly, and those reared artificially by Scheffler and Kerr (1975) reached a mean of about 2.6 kg at 1 year old compared with a mean of about 2.2 kg in females of the same age (Fig. 13.1). The mean weight of 62 yearlings artificially reared by Chamove (1981) was about 2.9 kg (Fig. 13.1). Weight gain during the first 50 days averaged about 9 g/d. Some data on growth rate have also been recorded by Harvey *et al.* (1979).

The growth of mother-reared female infants at a zoo and a laboratory (12 in all) was similar (Faucheux *et al.* 1978), these animals reached a mean weight of 2.35 kg at 1 year of age. At about 3 years of age the growth rate increases (Fig. 13.2) although the timing of this spurt and the rate of growth are dependent on the standard of management (Faucheux *et al.* 1978). Chamove (1981) recorded that although artificially reared infants grew more rapidly than those reared by their mothers in the early years, the growth of the mother-reared infants eventually surpassed them.

Milk and milk intake

Although it is likely that the dry matter content of the milk of *M. arctoides* exceeds that of the milk of man, and that the protein and fat form slightly higher, and the carbohydrate slightly lower, proportions of the dry matter, human milk replacers have been used very successfully to rear infants

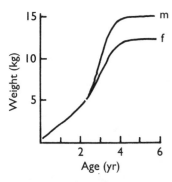

Fig. 13.2. Average growth curves of male (m) and female (f) stump-tailed macaques to maturity under good conditions. From Faucheux et al. (1978).

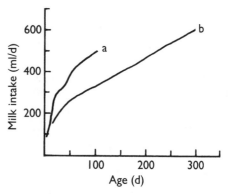

Fig. 13.3. Approximate daily milk intake of nursery-reared stump-tailed macaques. (a) from Chamove (1981), (b) from Scheffler and Kerr (1975).

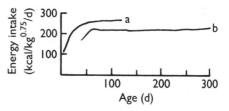

Fig. 13.4. Daily energy intake provided by milk per metabolic weight in nursery-reared stump-tailed macaques: (a) from Chamove (1981), (b) from Scheffler and Kerr (1975).

of this species. Chamove and Anderson (1982) reported the use of SMA supplemented with one drop of Abidec daily for rearing infants to 110 days of age. After this the diet was changed to cow's milk supplemented with dry chow and fresh fruit, and the amount of milk offered was gradually reduced.

On the first day of separation the average milk intake of Chamove and Anderson's (1982) groups of animals was 46–57 ml, rising to 120–125 ml the following day, and to 297–372 by 10 days of age. The daily intake gradually increased to a mean of about 500 ml/d by 110 days of age (Fig. 13.3). The intakes of those reared by Scheffler and Kerr (1975) were lower; from their graphs it appears that intake was about 175 ml at 20 days of age, rising to about 350 ml at 110 days of age and to 450 ml at 180 days (Fig. 13.3). From these data it would appear that daily intake in relation to metabolic weight rises to between 210 and 260 kcal/kg$^{0.75}$ by 60 days of age and remains quite constant during the first year (Fig. 13.4).

There is some evidence that self-feeding infants consume more than those mother-reared (Chamove and Anderson 1982). Overfeeding may occur and these authors cite one case of a baby which died of congestive heart failure at 9 weeks old having been consuming almost twice the average daily intake.

Lactation and weaning

The sucking pattern of infant stump-tailed macaques is similar to that of other macaque species (Roonwal and Mohnot 1977). Lactation has been observed to last between 6 and 18 months. Older siblings are finally prevented from sucking by the birth of the next baby, if not before (Trollope 1978).

Feeding

Chamove and Anderson (1982) described a technique by which baby stump-tailed macaques were trained to feed themselves within 40 hours. A tube feeder (designed for premature human infants) fitted with a small nipple was used and the authors considered it important to have quite a large hole in the nipple as the infants could not suck strongly at first. Once an hour between 09.00 h and 17.00 h (feeding later in the evening was found not to be necessary) each baby was lifted into position on the ramp, the nipple directed into its mouth, and 1–2 ml of SMA expressed. The milk was replaced 5 times daily and the babies allowed to feed *ad libitum*. Efforts were made to ensure that those reluctant to feed consumed at least 20 ml per day. Scheffler and Kerr (1975), whose interest was to study growth performance on a standardized diet, maintained their infants on Similac fed *ad libitum* to which a drop of Paladec was added daily. However, for the first 24 hours after feeding commenced (16 hours after parturition), a warm 10 per cent dextrose solution was offered every 2 hours. The following day, a 50:50 mixture of 10 per cent dextrose solution and Similac was provided, and thereafter Similac was given at full strength. The animals were hand-fed for about 10 days around which time they soon learned to feed themselves from a bottle attached to the side of the cage.

Accommodation

The animals reared by Scheffler and Kerr (1975) were housed in individual cages, with an ambient temperature of 24 °C, humidity 35 per cent, and dimmed light from 21.00 h to 07.00 h. Heating pads for extra warmth were provided for the first 15 days. Dim light is important at night to enable self-feeding babies to see the feeder (Chamove and Anderson 1982).

Until they were consistently feeding themselves Chamove (1981; see also Chamove and Anderson 1982) kept his animals in incubators at a temperature of 27–32 °C, in each of which was placed a towel-covered mesh ramp resting at an angle of 45 degrees, with a feeder clipped to the top. When they had learned to feed they were moved to heavy duty white polythene tubs measuring 0.58 × 0.65 × 0.75 m fitted with a wire mesh door. To prevent very young infants climbing the mesh a sheet of plastic was attached to the inside of the door. The temperature was held at 27–32 °C and a 40 watt light bulb attached to the outside of the cage provided a warmer area within it.

Infant management notes

When mothers did not entirely abandon their babies, but the standard of the care they provided was judged to be poor, Chamove (1981) left the infants with their mothers but gently restrained the mothers in a crush cage (in which a movable side is used for restraint by gentle pressure once daily and (whilst also feeding the mothers) fed the infants at least 30 ml of either SMA or Ionalyte. He found that these babies would often start feeding from their mothers after a few days. If this did not occur then the babies were taken for artificial rearing. Those babies that had been previously fed whilst still with their mothers learned to feed themselves sooner than those not previously fed in this way. Chamove (1981) considered that the best time to remove the infant, that is the least stressful to mother and baby, was 24 hours after birth. The longer the removal is delayed the greater the distress of separation and the longer the infant takes to feed itself. Infants can be removed after light chemical restraint (sedation) of the mothers using, for example, ketamine hydrochloride (Sainsbury *et al.* 1989).

Both Chamove (1981) Scheffler and Kerr (1975) initially housed babies individually all the time, but Chamove (1981) suggested that for normal development, social experience with at least one other animal for at least 2 hours a day before the age of 6 months is the minimum necessary and more social contact is desirable. From 90 days of age, Scheffler and Kerr (1975) provided social experience in groups for 4 hours per day.

Scheffler and Kerr (1975) provided a terrycloth-covered sandbag and cloth diapers for infants during the first 4 months and Chamove (1981) provided a towel-covered mesh ramp with a feeding bottle attached at the top (see Accommodation).

Physical development

The pattern of physical development is similar to that of the rhesus macaque (Chapter 12).

Behavioural development

Infants remain in close contact with their mothers for the first month, but then begin exploratory excursions (Roonwal and Mohnot 1977). They may continue to sleep with their mothers after the birth of a younger sibling (Trollope 1978).

Disease and mortality

Harvey and Rhine (1983) reported that 10 out of 64 babies were stillborn or died within two days of birth, and Bernstein and Gordon (1977) found that 4 (18 per cent) of 22 liveborn infants died within a month of birth. Hendrickx and Nelson (1971) did not provide any information on postnatal losses but in their study 26 (37 per cent) of 70 conceptions resulted in abortions, premature births or still births. Trollope and Jones (1975) reported only one stillborn out of 18 births, and of these 17 live babies only one died (at 6 months) prior to 2 years of age.

Of the 18 babies hand-reared by Scheffler and Kerr (1975) four died between 238 and 303 days post-partum, of an acute disease thought to be caused by a virus.

In captivity, the infant mortality rate of this species can be low. In the colony at the University of Stirling, 5 (6 per cent) of 83 were born dead, and only 5 (6 per cent) of the 78 liveborn died within 6 months of birth (3 of these were born to one mother who killed them and 2 died of hypothermia due to a heating fault (Chamove 1981). This author also described an outbreak of salmonellosis which killed two 3-year-olds, and mentioned that from 7 weeks of age the infants were sometimes observed to develop swollen lymph nodes in axillae and groin. He also noted that some individuals were prone to developing 'bloat' after feeding and one 3-year-old died as a result.

Preventative medicine
See notes for the rhesus macaque (Chapter 12).

Indications for hand-rearing
Captive-bred and, in particular, primiparous females have a tendency to inadequate parental behaviour (Chamove and Anderson 1982). Chamove (1981) provided supplementary feeding during the first few days if necessary until parental behaviour had improved (see Infant management notes).

Reintegration
See notes for the species macaque (Chapter 12).

References
Bernstein, I.S. and Gordon, T.P. (1977). Behavioural research in breeding colonies of old world monkeys. *Laboratory Animal Science*, **27**, 532–40.
Brandt, E.M. and Mitchell, G. (1971). Parturition in primates: behaviour related to birth. In *Primate behaviour: developments in field and laboratory research*, Vol. 2 (ed. L.A. Rosenblum), pp. 117–23. Academic Press, New York.
Brüggermann, S. and Grauwiler, J. (1972). Breeding results from an experimental colony of *Macaca arctoides*. In (Ed) *Breeding primates* (ed. W.I.B. Beveridge), pp. 216–26. Karger, Basel.
Bunyak, S.C., Harvey, N.C., Rhine, R.J., and Wilson, M.I. (1982). Venipuncture and vaginal swabbing in an enclosure occupied by a mixed-sex group of stump-tailed macaques (*Macaca arctoides*). *American Journal of Primatology*, **2**, 201–4.
Chamove, A.S. (1981). Establishment of a breeding colony of stumptailed monkeys (*Macaca arctoides*). *Laboratory Animals*, **15**, 251–9.
Chamove, A.S. and Anderson, J.R. (1982). Hand-rearing infant stump-tailed macaques. *Zoo Biology*, **1**, 323–31.
Dukelow, W.R. (1973). Captive breeding of non-human primates. *Proceedings of the American Association of Zoo Veterinarians*, 52.
Estrada, A. and Estrada, B. (1976). Birth and breeding cycle in an outdoor living stumptail macaque (*Macaca arctoides*) group. *Primates*, **17**, 225–31.
Faucheux, B., Bertrand, M., and Bourliere, F. (1978). Some effects of living conditions on the pattern of growth in the stumptail macaque (*Macaca arctoides*). *Folia Primatologica*, **30**, 220–36.
Harvey, N.C. and Rhine, R.J. (1983). Some reproductive parameters of stump-tailed macaques (*Macaca arctoides*). *Primates*, **24**, 530–6.
Harvey, N.C., Rhine, R.J., and Bunyak, S.C. (1979). Weights and heights of stump-tailed macaques (*Macaca arctoides*) living in colony groups. *Journal of Medical Primatology*, **8**, 372–6.
Hendrickx, A.G. and Giles Nelson, V. (1971). Reproductive failure. In *Comparative reproduction of non-human primates* (ed. E.S.E. Hafez), pp. 403–25. Charles C. Thomas, Springfield, Illinois.
IUCN (1990). *1990 IUCN red list of threatened animals*. IUCN, Gland, Switzerland.
Johnsen, D.O. and Whitehair, L.A. (1986). Research facility breeding. In *Primates. The road to self sustaining populations* (ed. K. Benirschke), pp. 499–511. Springer-Verlag, New York.
Martin, D.P. (1986). Primates: reproduction and obstetrics. In *Zoo and wild animal medicine* (ed. M.E. Fowler), pp. 701–4. W.B. Saunders, Philadelphia.
McCann, C. (1933). Notes on some Indian macaques. *Journal of the Bombay Natural History Society*, **36**, 796–10.
Michael, R.P. and Zumpe, D. (1971). Patterns of reproductive behaviour. In *Comparative reproduction of non-human primates* (ed. E.S.E. Hafez), pp. 205–42. Charles C. Thomas, Springfield, Illinois.
Roonwal, M.L. and Mohnot, S.M. (1977). *Primates of South Asia*, pp. 97–171. Harvard University Press, Cambridge, Mass.
Sainsbury, A.W., Eaton, B.D., and Cooper, J.E. (1989). Re-

straint and anaesthesia in primates. *Veterinary Record*, **125**, 640–3.

Scheffler, G. and Kerr, G.R. (1975). Growth and development of infant *M. arctoides* fed a standardised diet. *Journal of Medical Primatology*, **4**, 32–44.

Stenger, V.G. (1972). Studies on reproduction in the stump-tailed macaque. In *Breeding primates* (ed. W.I.B. Beveridge), pp. 100–4. Karger, Basel.

Trollope, J. (1978). Reproduction in a closed colony of *Macaca arctoides*. In *Recent advances in primatology*, Vol. 2 *Conservation* (ed. D.J. Chivers and W. Lane-Petter), pp. 243–50. Academic Press, London.

Trollope, J. and Jones, N.G.B. (1975). Age of sexual maturity in the stump-tailed macaque (*M. arctoides*): a birth from laboratory parents. *Primates*, **13**, 229–30.

Trollope, J. and Jones, N.G.B. (1975). Aspects of reproduction and reproductive behaviour in *Macaca arctoides*. *Primates*, **16**, 191–205.

Trum, B.F. (1972). Research on reproduction at the New England Regional Primate Research Center. In *Breeding primates* (ed. W.I.B. Beveridge), pp. 198–202. Karger, Basel.

Wolfheim, J.H. (1983). *Primates of the world*, pp. 464–8. University of Washington Press, Seattle.

Gelada baboon

14 Common baboon

Species

The common baboon *Papio cynocephalus*

ISIS No. 140600800400

Status, subspecies, and distribution

There is some debate concerning the classification of baboons. The common or savanna baboon (also known as the yellow or olive baboon) is widely distributed in the African continent, and inhabits savanna, woodland, and the forest edge. This species is not threatened at present, but the Hamadryas baboon *Papio hamadryas* and the Gelada baboon *Theropithecus gelada* are both listed as rare (Lee *et al.* 1988).

The baboon has been widely used in biomedical research and it was estimated that there were 3225 potential and actual breeding specimens in US principal research institutions in 1984 (Johnsen and Whitehair 1986). In the Soviet Union, a tenth of all primate experiments involve baboons (Fridman and Popova 1988).

Sex ratio

The sex ratio at birth is probably close to 1:1 (Altman and Altmann 1970).

Social structure

In the wild, baboons live in troops ranging in size from about 15 to 200 individuals (Altmann and Altmann 1970; Michael and Zumpe 1971). Related females tend to remain within the social group, whereas males move to other groups at adolescence (Smuts 1985).

In captivity established groups are resistant to the introduction of new individuals. New social groups have been established by simultaneous introduction of 20 to 40 individuals into a large social group cage (Goodwin and Coelho 1982). After a period of two to eight months preliminary stabilization the new groups were introduced to an approximately circular corral of 2.5 hectares designed to hold up to 600 adult baboons. Using this technique Goodwin and Coelho (1982) introduced 427 baboons to the corral with no deaths.

Breeding age

In the female baboon, puberty is reached at the age of 3–4 years of age (Hafez 1971) or even as late as 5 years (Richard 1985). From the time of puberty, the sexual skin increases in size with successive cycles for one to two years (Hafez 1971). Breeding occurs one year (Richard 1985) or up to two years after puberty, and physical maturity and full adult weight are reached at two to three years after puberty (Smuts 1985). In captivity, sexual maturation occurs at an earlier age, probably because of a higher level of nutrition. In the wild, the age at first reproduction is dependent on food availability during growth (Richard 1985). Menstruation and breeding may continue for up to 20 years of age (Newsome 1967; Hafez 1971; Richard 1985).

In the male, active spermatogenesis begins at 3 years of age (Hafez 1971) but, in the wild, breeding does not usually occur until later due to social constraints. Males reach full physical maturity at 9–10 years of age (Smuts 1985).

Longevity

Maximum lifespan is over 30 years (Newsome 1967; Jones 1968).

Seasonality

Baboons are not strictly seasonal breeders (Altmann 1980); births occur throughout the year although there may be seasonal peaks in some populations (Smuts 1985), and peaks of fertility may also be observed in captive-bred colonies (Hearn 1984). Reports on the length of menstrual cycles range from 31 to 35 days (Michael and Zumpe 1971) to an average of 41 days (Miller and Pallota 1965) although it has been observed that cycles may vary considerably, particularly in captivity (Hafez 1971). A cycle is characterized by the swelling of the sexual skin on the rump of the female. This period of increasing turgescence lasts three to nine days (Michael and Zumpe 1971) and results in the perineal skin appearing glossy red or bright pink, smooth, and swollen (Hendrickx and Kriewaldt 1967; Hendrickx 1967). This swelling remains at a peak for approximately 10 days followed by an abrupt detumescence lasting eight days (Miller and Pallota 1965). The quiescent phase is characterized by a dry, dull, wrinkled white/pink appearance of the perineal area (Hendrickx and Kriewaldt 1967) which lasts approximately two weeks, (Miller and Pallota, 1965) or 10 days (Smuts 1985). Ovulation occurs two to three days (Michael and Zumpe, 1971) or up to four days (Smuts 1985) before the onset of detumescence. Menstruation begins in the latter part of detumescence (Miller and Pallota 1965) or during the quiescent phase before the sexual skin begins to swell again (Michael and Zumpe 1971; Smuts 1985). Vaginal bleeding is observed for three to four days (Miller and Pallota 1965; Michael and Zumpe 1971) but this is not always overt in some individuals and vaginal smears are sometimes necessary to detect it (Hendrickx and Kriewaldt 1967). A post-partum amenorrhoea is observed in the baboon which lasts from six to 16 months (Richard 1985) or even up to 21 months (Smuts 1985). A further report estimates this lactational amenorrhoea to average 420 days but is followed by 189 days of cyclic activity, before conception takes place (Hearn 1984) and Altmann (1980) reports a 12 month post-partum amenorrhoea in *Papio* species in the field followed by an average of four cycles before conception. Therefore, interbirth intervals in the baboon which vary greatly among individuals (Altmann 1980) may be very long, such as 18 to 24 months (Hearn 1984) or up to 26.5 months (Smuts 1985), although in Uganda interbirth intervals tend to be just over a year (Rowell *et al.* 1968). If the infant is stillborn or dies, the post-partum amenorrhoea is curtailed and the interbirth interval is shortened. In this case, a female may cycle within a month of her infant's death, and an average of two cycles follow before conception (Altmann 1980). This interval may also depend on the supply of food and the age of the female: it is shortest at eight years of age but longer if food is scarce or if the female is very young or over 16 years of age (Richard 1985).

Gestation

Reports of gestation length include 154–183 days (Napier and Napier 1967), 173–193 days (Michael and Zumpe 1971), 173 ± 2 days (Buss and Voss 1971) and a mean of 183.4 days (Nowell *et al.* 1978). Dubouch (1969) distinguishes between *P. cynocephalus* with a gestation period of 166 days) and *P. anubis* (gestation 185 days).

Pregnancy diagnosis

Diagnosis of pregnancy may be accomplished by digital palpation of the uterus per rectum to detect an increase in uterine size (Hendrickx and Kriewaldt 1967; Max Lang 1971). With experience, pregnancy can be detected by this method as early as 14 days following conception (Hendrickx and Kriewaldt 1967). Visual examination of the sexual skin is also useful: 10 days post-coitus, the skin of a non-impregnated female will be white, rough, and scaly but if pregnant, the skin is pink or red and this colour remains until after parturition (Hendrickx and Kriewaldt 1967; Richard 1985).

A test based on detection of urinary chorionic gonadotrophin can be used for pregnancy diagnosis in early gestation but not in the last weeks before parturition (Hodgen and Nieman 1975; Mack and Kafka 1978).

In the later stages of pregnancy fetal bones can be seen on radiography of fluoroscopy (Love 1978). Ultrasonography has also been used to evaluate pregnancy (Farine *et al.* 1988). An increase in the size of the abdomen is noticeable at three to four months of gestation (Newsome 1967).

Birth

A high incidence of breech births, 12 to 16 per cent, was recorded by Max Lang (1971). These may be detected and sometimes corrected by rectal palpation. Birth takes place at night (Miller and Pallota 1965; Kalter 1977), usually between

22.00 h and 06.00 h (Beattie 1972). The female cleans the infant, removes the placenta and severs the umbilical cord (Moore and Cummins 1979). Most baboons (69 per cent) eat the placenta (Max Lang 1971). One observed delivery lasted 18 minutes and the female assisted in the birth by pulling on the infant. The placenta was delivered almost immediately and the umbilical cord remained unsevered for 5 hours (Beattie 1972). In one *Papio anubis* (Love 1978), the first stage of labour lasted an unusually long time, with a marked loss of appetite by the female and a loss of fluids from her vulva during the last day or so before parturition. During the second stage of labour, the female usually squatted during the abdominal contractions and pulled on the bars of her cage with her hands. In between contractions, she moved around her cage. Once the infant's head appeared, the mother cleaned away the membranes and pulled on the infant at the next contraction so that the whole infant was delivered. When the placenta was delivered, the female spent 25 minutes consuming it up to the umbilical cord during which time she neglected the infant. Having done this, she relaxed and cradled her infant. In another report on a birth in *Papio anubis* (Nowell *et al.* 1978), a small amount of bleeding from the vagina was noticed 71 minutes before birth. Abdominal contractions only began 42 minutes before parturition when the female assumed several positions while arching her back, extending her limbs, twitching her tail, and frequently touching or licking the perineal area. During delivery, which was rapid, she stood with her rump raised. The infant vocalized within 20 seconds of birth and was immediately cradled and groomed and the infant was clinging to the mother's ventrum when she walked away four minutes later. It has been reported that turning from the breech to the cephalic presentation in baboons may occur as late as during the last 24 hours before delivery (Max Lang 1971).

Litter size
Usually the litter size is 1 (Michael and Zumpe 1971; Richard 1985) but twins have occasionally been recorded (Hendrickx and Giles Nelson 1971).

Adult weight
Male baboons weigh 20 to 30 kg when adult, and adult females weigh 11 to 15 kg (Whitney and Wickings 1986).

Neonate weight
Male babies weigh between 798 and over 1000 g, and female babies weigh between 760 and 830 g (Miller and Pallota 1965; Creager and Switzer 1967; Buss and Voss 1971; Moore and Cummins 1979). Vice *et al.* (1966) reported a neonate weight range of 539–1077 g. The greatest neonate weights recorded have been of captive-bred babies. A premature infant delivered by Caesarean section at 159 ± 2 days of gestation weighed 575 g (Chorazyha 1972).

Adult diet
In the wild, *P. cynocephalus* feed on a wide variety of foods. Green grass, rhizomes, and the base of blades when grass is dry are regularly eaten. Acacia trees also provide leaves, blossoms, seeds, pods, sap or gum. The leaves, fruits, and flowers of bushes, herbs, bulbs, and roots are also taken and small mammals and even infant ungulates are sometimes hunted and eaten (S.A. Altmann and J. Altmann 1970; Smuts 1985). In captivity, baboons can be maintained on commercial diets, such as Purina Monkey Chow (Miller and Pallota 1965; Dollahite Wene *et al.* 1982) or SDS Mazuri pellets. Consumption of these diets approximates 220–300 g per animal daily (Dollahite Wene *et al.* 1982). Supplements of fruits and vegetables are important (Miller and Pallota 1965), particularly for nursing mothers (Beattie 1972). Baboons show individual preferences for different flavours, and may become obese if fed a preferred food *ad libitum* (Dollahite Wene *et al.* 1982).

Adult energy requirements
The daily energy requirement of baboons is 64 to 80 kcal/kg (Buss and Voss 1971), or about 120 to 150 kcal/d per $kg^{0.75}$.

Growth
Female baboons in the wild reach their full adult weight at 6 to 8 years of age whereas males reach

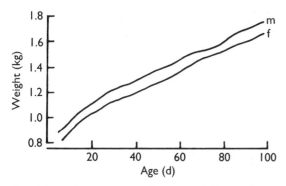

Fig. 14.1. Average growth curve of hand-reared infant male (m) and female (f) baboons. From Moore and Cummins (1979).

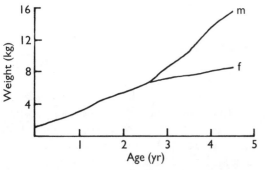

Fig. 14.2. Average growth curves of male (m) and female (f) Papio anubis, Papio cynocephalus and their hybrids. From Sigg et al. (1982).

their full weight at 9 to 10 years (Smuts 1985). When infants are removed for hand-rearing at birth, some weight loss may occur (3–15 g) during the first 3 days of life (Moore and Cummins 1979).

Figure 14.1 shows the growth of 9 male and 15 female infants reared on a diet providing 0.7 kcal/ml from birth to 110 days (Moore and Cummins 1979). These animals grew at an average rate of 7 g/d. Data for one male and two females hand-reared for 40 days by Creager and Switzer (1967) are similar. Figure 14.2 indicates the growth male and female baboons (*Papio anubis*, *Papio cynocephalus*, and their hybrids) reared in the laboratory from birth to 4.5 years of age (Sigg et al. 1982). *Papio ursinus* reared on a semi-synthetic diet gained weight at a rate of 15 g/d throughout their first year (Du Bruyn and De Klerk 1978), reaching an average body weight of about 6 kg at 1 year old (see also Snow 1967).

Milk and milk intake

The average composition of milk from *Papio* species has been found to be 4.6 per cent fat, 1.5 per cent protein, 7.7 per cent carbohydrate, and 0.3 per cent ash (Hummer 1970; Buss 1968, 1971). On a dry matter basis, the composition approximates 33 per cent fat, 9 per cent protein, 55 per cent carbohydrate, and 2 per cent ash (Oftedal 1984). The energy density of the fresh milk is 0.8 kcal/ml (Buss 1971; Buss and Voss 1971) and the milk contains 14.3 g solids per 100 ml (Buss 1968). The concentrations of some major elements have also been measured as follows: calcium 0.4 g/l, chloride 0.2 g/l, potassium 0.4 g/l, phosphorus 0.3 g/l, and sodium 0.1 g/l (Buss 1971).

The fat in baboon milk is present as small globules which slowly separate on standing. The fat concentration increases as lactation progresses. Within the protein fraction, there is twice the amount of casein compared to whey protein. Large amounts of lysozyme are present which are thought to protect the mammary glands from mastitis. The colostrum differs in composition from the later milk and contains less lactose and fat and more protein and ash. It is also relatively poor in casein but rich in immunoglobulin type A (Buss 1971). The specific gravity of milk produced between days 11 and 35 of lactation is 1.027 and its pH is 7.44 (Buss 1968).

The composition of baboon milk is similar to that of human milk and human milk replacers such as Similac, Similac-with Iron, SMA and Enfamil have generally been used to feed infant baboons (Miller and Pallota 1965, Buss and Voss 1971; Moore and Cummins 1979; Creager and Switzer 1967; Griffin et al. 1986). These have mainly been used at the concentration recommended for human infants but some have used higher concentrations (Vice et al. 1966; Kalter 1977). Vice et al. (1966) found that higher energy intakes and greater weight gains, with resulting better health, occurred when Similac was made up to provide 0.87 kcal/ml than when made up to provide 0.43 kcal/ml. Similarly Moore and Cummins (1979) observed that diets with greater energy density (about 0.7 kcal/ml) resulted in better growth and health and fewer clinical problems than diets with a lower energy concentration.

The composition and energy density of human milk replacers can be made to mimic baboon milk more closely by adding Mazola (corn) oil to raise the fat content to 4.8 g/ml, and this was recommended by Buss *et al.* (1970). The energy density of the milk is then 0.8 kcal/ml. It is better to increase the energy density by adding unsaturated fatty acids in this way than by increasing the concentration of the human formula. Vitamin and mineral supplements have been used (Miller and Pallota 1965; Creager and Switzer 1967). Some recommend supplementation with a human infant vitamin preparation, but deficiencies are perhaps unlikely using modern human milk replacers. The formula may be replaced by cow's milk at 4 to 5 months of age (Miller and Pallota 1965) and infants may be completely weaned by 4 to 6 months of age (Martin 1986).

Before the composition of baboon milk had been measured, a formula based on half-skimmed, sweetened, powdered cow's milk was used by Dubouch (1969) to hand-rear baboons and this appeared to support a growth rate greater than observed in mother-reared babies. Du Bruyn and De Klerk (1975) fed infants of less than 5 weeks old that were captured from the wild on a semi-synthetic diet based on casein, sunflower oil, agar, sugar, dextrin, and mineral and vitamin compounds.

Infants that suckle from their mothers are estimated to consume up to 200 ml of milk daily during their first week of life (Buss 1971), 250 ml daily at 2 weeks and 350 ml/day by females or 450 ml/day by males at 4 months of age (Buss and Voss 1971). The maximum intake relative to body weight occurs at 2 weeks of age (Creager and Switzer 1967) or between 2 and 6 weeks of age (Buss and Voss 1971). However, hand-reared animals fed *ad libitum*, particularly those taught to self-feed, will consume larger amounts than those infants raised naturally and thus growth rates are higher as a result. It has been observed that infants can take up to three times the volume of milk that maternally nursed infants are estimated to consume. Small changes in the caloric content of milk do not appear to affect milk intake (Buss and Voss 1971).

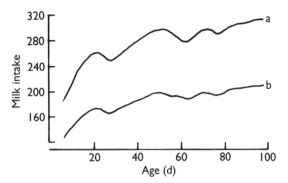

Fig. 14.3. Approximate daily (a) milk, ml/d, and (b) energy intake, kcal/d provided by milk in hand-reared baboons. From Moore and Cummins (1979).

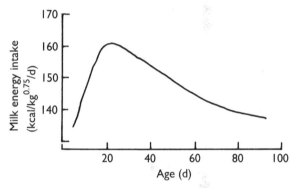

Fig. 14.4. Approximate daily energy intake provided by milk in hand-reared male baboons per metabolic weight in relation to age. From Moore and Cummins (1979).

The mean daily milk formula consumption and energy intake of nine male infant baboons during their first 100 days (Moore and Cummins 1979) is shown in Fig. 14.3. The energy intake in relation to metabolic weight increased rapidly to about 160 kcal/d per unit metabolic weight, then gradually declined to about 140 kcal/d (Fig. 14.4). Male infants consume more than females and their weight increases slightly more rapidly (Creager and Switzer 1967; Du Bruyn and De Klerk 1978).

Lactation and weaning

The infant nurses within a few minutes of birth and colostrum is produced for up to 24 hours post-partum (Buss 1971). During the first six months, 10 to 20 suckling bouts, of short duration, are observed hourly. The frequency of suckling decreases over the next six months from seven bouts per hour to two bouts at 12 months of age (Hearn 1984). The milk yield increases during the first four months of lactation (Buss and Voss 1971) from up to 200 ml/day at 1 week to 400 ml/day at 4 months (Buss 1971). *Papio cynocephalus* infants mouth solid food at approximately 1 month of age but significant amounts are not taken until 2 to 3 months (Buss 1971), although Chance and Jolly (1970) observed infants taking their first solids at 4 to 6 months of age. It was suggested that the exact timing may depend on the availability of accessible weaning foods, such as flowers fallen to the ground from trees (for example, umbrella and fever trees in Kenya) (Altmann 1980). Infants are no longer nursing regularly by the end of the first year (Buss 1971; Chance and Jolly 1970) but the weaning process continues over into the second year when the mother increasingly rejects her infant if it tries to ride on her back or take the nipple (Chance and Jolly 1970). However, suckling may be observed up to 2 years of age (Smuts 1985), although this may be for security rather than milk (Buss 1971).

Semi-solid food, such as strained foods, may be introduced very soon after birth (Kalter 1977). A mush of commercial chow biscuits soaked in water and apple sauce may be offered in a bowl. Young infants will play with and mouth such a mash. Caution has been voiced about offering such foods before 18 days of age as they provide a good medium for proliferation of *Salmonella* (Martin 1986). Pablum and/or soaked biscuits were introduced at 3 weeks of age by Hummer (1970), but in other reports solids were not introduced until later. Baboon biscuits were offered at 14 weeks of age (Moore and Cummins 1979), or at 10 to 12 weeks of age in another report (Buss and Voss 1971). Farex cereal for human babies was offered to infants removed from their mothers at 10 weeks of age but was not readily taken (Baker *et al*. 1978). Miller and Pallota (1965) did not introduce solids (monkey chow, fruits, and baby food) until the age of 4 to 5 months (Miller and Pallota 1965). The infants reared by Griffin *et al*. (1986) were, in general, maintained on an adult diet by 6 months of age.

Feeding

When infant baboons are separated from their mothers at birth, it has been recommended that the first feeds consist of a hypotonic solution followed by 5 per cent dextrose (Miller and Pallota 1965) or 5 per cent dextrose for the first two days (Hummer 1970) or a 1:1 mixture of 5 per cent dextrose and milk formula (Martin 1986). In one report (Moore and Cummins 1979) the infants were separated at 6 to 12 hours after birth having had a chance to obtain colostrum, and were on undiluted milk replacer straight away. There may be little advantage in using 5–10 per cent dextrose or glucose solutions as opposed to milk replacer if the baby sucks well and there is little danger of aspiration.

The use of 120 ml feeding bottles was recommended by Moore and Cummins (1979) and the standard pre-packaged human infant nipples were adequate, although they needed softening by boiling in water and slight enlargement of the nipple holes.

Infant baboons can be trained to self-feed from a suspended bottle by 1 week of age (Vice *et al*. 1966; Kalter 1977) but hand-feeding for the first month allows accurate monitoring of the daily intake (Moore and Cummins 1979).

Feeding schedules vary greatly among the published hand-rearing reports but it is generally agreed that night feeds (between 24.00 h and 08.00 h, for example) are not necessary (Moore and Cummins 1979; Martin 1986) unless the infant is ill, weak, or premature (for example, Chorazyna 1972). Five daily feeds or up to seven daily feeds are recommended for the first two to three weeks of life (Hummer 1970; Buss and Voss 1971; Moore and Cummins 1979; Griffin *et al*. 1986). The frequency of feeding should be gradually reduced to three daily feeds at 4 weeks of age (Moore and Cummins 1979) or 6 weeks of age (Buss and Voss 1971) according to solid food intake.

Accommodation

Newborn infants should be kept in incubators until their body temperatures have stabilized, a process that usually requires two to three days. Ambient temperature should be maintained at 31–33 °C so that the infants' rectal temperature is maintained at 37 °C (Moore and Cummins 1979).

Infants may be kept in incubators for the first one or two months of life and then moved into cages in groups of five animals (Miller and Pallota, 1965). It has been suggested, however, that infants need remain in incubators for only 10 to 14 days before being moved into wire cages that provide more space and are more easily cleaned (Moore and Cummins 1979).

Soft mother-surrogates or towels should be provided for the infant to cling to. Infant baboons that are reared in isolation from others exhibit abnormal behaviour, including listlessness, curling up at the bottom of the cage, and sucking thumbs, toes, or penis (Moore and Cummins 1979).

Infant management notes

Newborn infants taken for hand-rearing should be dried with a sterile towel, or if necessary, cleaned in a warm bath with dilute Betadine surgical soap and then dried. The umbilical cord should be ligated with silk and severed leaving a length of 3 cm. The end should be treated with iodine (Moore and Cummins 1979).

Hygiene is important (Moore and Cummins 1979; Martin 1986). Incubators should be cleaned and disinfected daily and wire cages should be cleaned weekly (Moore and Cummins 1979). Diapers placed in the incubator should be changed after each feed (Moore and Cummins 1979).

Daily records of food consumption, rectal temperature, activity level, body weight, and character of the faeces should be kept (Moore and Cummins 1979). These authors consider that 'burping' after feeding is important, particularly for the first week of life. This should be carried out as for human babies once the normal amount of food has been consumed as premature burping will make the infant lose interest in sucking (Miller and Pallota 1965; Moore and Cummins 1979). If necessary fluids or milk can be administered by intra-gastric tube (Miller and Pallota, 1965).

Physical development

Papio species are born fully furred and with their eyes open. The natal coat is relatively dark (Smuts 1985) or black (Altmann 1980) and the skin is pink but changes to grey from the third or fourth month (Napier and Napier 1967; Rowell *et al.* 1968; Altmann 1980). A degree of pigmentation appears in the hands and feet at this time as do some gold hairs in the coat, particularly around the wrists and in the brow region (Altmann 1980). The natal coat is replaced by the adult pelage at 6 to 7 months (Altmann 1980; Smuts 1985) and the change to the full adult skin and pelage colour is complete by 10 to 12 months (Napier & Napier 1967), although a little pink skin in the muzzle and ears may be present by the end of the first year and the scrotum of the male remains pink until the second birthday (Altmann 1980).

In a premature infant delivered by Caesarean section, the first tooth appeared by the end of the third week instead of during the first week. All eight deciduous incisors were visible at 7 weeks of age and the first four molars appeared at 13 weeks (Chorazyha 1972). The eruption of the permanent dentition occurs as follows: centrals incisors 33–35 months, lateral incisors 36–38 months, canines 48–49 months, first premolars 49–50 months, second premolars 55 months, first molars 19–24 months, and second molars 45–49 months. There are no data on the eruption of the third molars (Reed 1965).

The testes in the male descend at puberty (Kinzey 1971); between 4.8 and 6.8 years of age (Sigg *et al.* 1982).

Behavioural development

Infant *Papio* species are born with good clasping and grasping reflexes and cling on to their mothers shortly after birth (Nowell *et al.* 1978). Initially, they cling to the mother's ventrum but later ride on her back (Smuts 1985). They sometimes need support in the first few days during troop movements when the female will reposition her infant and may even have to walk three-legged (Altmann 1980). At birth, the locomotor abilities are limited and for the first few weeks the limbs remain flexed and the gait is wobbly. Infants pick up objects with their mouths rather than their hands (Altmann 1980). At this stage, the infant spends a third of the day asleep on its mother with a nipple in the mouth (Rowell *et al.* 1968). The first break in contact between mother and infant is initiated by the infant and may occur as early as two weeks. During the first month the infant becomes more active and by 1 month of age it can move competently over the ground (Altmann 1980). Contact with its mother begins to decline at 5 weeks of age (Rowell *et al.* 1968). By the second month, it is the mother that initiates breaks in contact (Altmann 1980). At this time the infant starts climbing over obstacles and begins to manipulate objects with its hands (Altmann 1980). By the end of the second month, the infant spends a significant amount of time out of reach of the mother (Rowell *et al.* 1968). During the third and fourth months, the infant spends increasing time playing with its peers (Altmann 1980). Rudimentary play first appears in the fourth week and by 3 months of age, the infant spends half its waking time involved in social play with other infants (Rowell *et al.* 1968). By the fourth month it stays away from its mother's ventrum while she feeds and starts to test accessible foods and learns to climb around in trees. At this stage, an infant is still dependent on its mother for food and transport. By four to six months, the infant experiences some rejection by its mother, and may exhibit 'tantrums' as a consequence (Altmann 1980). By this stage most infants ride on their mothers' backs during troop movements, sitting 'jockey-style' close to the female's tail or lying flat when at a run (Chance and Jolly 1970). Some infants have been observed riding dorsally as early as two months and others as late as eight months (Altmann 1980). Infants continue developing their tree climbing abilities and can usually descend without assistance by the age of eight months (Altmann 1980), at which stage they spend nearly all their waking time with other infants (Rowell *et al.* 1968). During months 9 to 12, dominance relationships appear among infants, and at this stage they sleep with their mothers but are no longer tolerated on their backs. By 1 year of age, infants are independent, although they may still take their mother's nipples in the mouth in times of stress or they may sleep on their mother's ventrum if she has no younger offspring (Altmann 1980).

Male baboons show complete copulatory behaviour in their first year (Rowell *et al.* 1968).

Disease and mortality

In the wild, mortality rates may be high for infants in their first two years of life. In one report (Altmann 1980), mortality was 28 per cent in the first year of life, 25 per cent in the second year, but zero in the next three years of life. Females provide little care for their offspring in the second year. Female infants appeared to show equal or higher rates of survival than males (Altmann 1980).

During the first two years after establishing a large-scale baboon breeding corral, Goodwin and Coelho (1982) reported 437 births. Of these 37 (8.5 per cent) were stillborn or died within the first 24 hours, and 72 (18 per cent of those surviving more than 24 hours) died at later stages of growth. The causes of mortality were infant-stealing by mature females, juveniles, and adult males, trauma, and diarrhoea and pneumonia.

Mortality rates among hand-reared infant baboons have been recorded at less than 5 per cent, and the major causes of death were pneumonia and birth anomalies (atresia ani, hydrocephalus, blindness). Intestinal problems were rarely fatal if treated (Moore and Cummins 1979). Within a small sample of hand-reared infant baboons, it was observed that males showed higher rates of survival perhaps due to their greater birth weights (Creager and Switzer 1967). Female baboons stressed before parturition, for example by transport, may kill or reject their infants (Moore and Cummins 1979).

Preventative medicine

Infants taken for hand-rearing should be given a complete physical examination. It may be helpful to record heart rate, heart sounds, respiratory rate, body weight, temperature, and the colour of mucous membranes. Some have advocated haematological screening at monthly intervals and the regular collection of oral and faecal swabs for microbiological examination. Pneumonia and enteritis are the major clinical problems of infant baboons (Moore and Cummins 1979).

Where tuberculosis is prevalent three intradermal tests at 30 day intervals and repeated every two to four months has been suggested (Hummer 1967). Annual chest radiographs may also aid in the detection of tuberculosis (Ott-Joslin 1986). Yellow fever vaccination is recommended for animals in or from high-risk areas (Hummer 1967; Ott-Joslin 1986). Vaccination against measles should be considered for newly imported baboons and those born in captivity.

Indications for hand-rearing

In the wild, poor mothering rarely results in the infant's death (Altmann 1980). The care provided by primiparous mothers was inferior to that given by experienced mothers, but it visibly improved within the first few days of birth (Altmann, 1980).

Hand-rearing is indicated if the mother is unwilling or too ill to rear the baby, or if the baby is too weak or ill to suck. It appears that infants of *Papio* species cannot be fostered on to other females (Buss 1971). However, because of the great advantages of this to the normal development of the infant, if an opportunity arises for cross-fostering an attempt should be considered.

Reintegration

Hand-reared infant baboons housed alone develop abnormal behaviours or neuroses. They lack the parental or social discipline necessary to develop and maintain well-adjusted responses to their environment in later life (Beattie 1972; Baker *et al.* 1978). In one report, it was recommended that hand-reared infants should be placed in activity cages with their peers for 2 hours daily starting from 1 week of age (Moore and Cummins 1979). In the absence of specific information for baboons, the principles of management of hand-reared rhesus macaques to ensure socialization for subsequent breeding success should be considered.

References

Altmann, J. (1980). *Baboon mothers and infants.* Harvard University Press, Cambridge, Mass.

Altmann, S.A. and Altmann, J. (1970). *Baboon ecology.* University of Chicago Press.

Baker, B.A., Morris, G.F., and Cowan, T.D. (1978). Breeding baboons in Uganda and Cambridge. In *Recent advances in primatology*, Vol. 2. *Conservation* (ed. D.J Chivers, and W. Lane-Petter), pp. 217–29. Academic Press, London.

Beattie, I.A. (1972). Some problems of breeding baboons under laboratory conditions. In *Breeding primates* (ed. W.I.B. Beveridge), pp. 48–54. Karger, Basel.

Buss, D.H. (1968). Gross composition and variation of the components of baboon milk during artificially stimulated lactation. *Journal of Nutrition*, **96**, 427–32.

Buss, D.H. (1971). Mammary glands and lactation. In *Comparative reproduction of nonhuman primates* (ed. E.S.E. Hafez), pp. 315–33. Charles C. Thomas, Springfield, Illinois.

Buss, D.H. and Voss, W.R. (1971). Evaluation of four methods for estimating the milk yield of baboons. *Journal of Nutrition*, **101**, 901–10.

Buss, D.H., Voss, W.R., and Nora, A.H. (1970). A brief note on a modified formula for hand-rearing infant baboons. *International Zoo Yearbook*, **10**, 133–4.

Chance, M.R.A. and Jolly, C.J. (1970). *Social groups of monkeys, apes and men.* Jonathan Cape, London.

Chorazyna, H. (1972). The rearing of a premature baboon. In *Breeding primates* (ed. W.I.B. Beveridge), pp. 55–7. Karger Basel.

Creager, J.G. and Switzer, J.W. (1967). Some factors relating to growth and development of primate infants with special reference to the baboon. In *The baboon in medical research*, Vol. 2 (ed. H. Vagteborg), pp. 85–98. University of Texas Press, Austin.

Dollahite Wene, J., Barnwell, G.M., and Mitchell, D.S. (1982). Flavor preferences, food intake, and weight gain in baboons (*Papio* sp.). *Physiology and Behaviour*, **28**, 569–73.

Dubouch, P. (1969). Artificial rearing of baboons. In *Primates in medicine*, Vol. 2 (ed. W.I.B. Beveridge), pp. 96–9. Karger, Basel.

Du Bruyn, D.B. and De Klerk, W.A. (1975). Feeding and rearing of infant baboons in captivity. *South African Medical Journal*, **49**, 2229–32.

Du Bruyn, D.B. and De Klerk, W.A. (1978). A semi-synthetic diet for growing baboons. *Journal of the South African Veterinary Association*, **49**, 193–5.

Farine, D., MacCarter, G.D., Timor-Tritch, I.E., Yeh, M.-N., and Stark, R.I. (1988). Real-time evaluation of baboon pregnancy: biometric measurements. *Journal of Medical Primatology*, **17**, 215–21.

Fridman, E.P. and Popova, V.N. (1988). Species of the genus *Papio* (Cercopithecidae, Primates) as subjects of biomedical research. II. Quantitative characteristics of contemporary use of baboon species in medical and biological investigations. *Journal of Medical Primatology*, **17**, 309–18.

Goodwin, W.J. and Coelho, A.M. Jr. (1982). Development of a large scale baboon breeding program. *Laboratory Animal Science*, **32**, 672–6.

Griffin, C.L., Musselman, R.P., Yeates, D.B., Raju, T.N., Harshbarger, R.D., and Lourenco, R.V. (1986). Hand-rearing baboons for laboratory investigations. *Laboratory Animal Science*, **36**, 686–90.

Hafez, E.S.E. (1971). Reproductive cycles. In *Comparative reproduction of non-human primates* (ed. E.S.E. Hafez), pp. 85–114. Charles C. Thomas, Springfield, Illinois.

Hearn, J.P. (1984). Lactation and reproduction in non-human primates. *Symposia of the Zoological Society of London*, **51**, 327–35.

Hendrickx, A.G. (1967). The menstrual cycle of the baboon as determined by the vaginal smear, vaginal biopsy and perineal swelling. In *The baboon in medical research*, Vol. 2, (ed. H. Vagteborg), pp. 437–59. University of Texas Press, Austin.

Hendrickx, A.G. and Giles Nelson, V. (1971). Reproductive failure. In *Comparative reproduction of non-human primates* (ed. E.S.E. Hafez), pp. 403–25. Charles C. Thomas, Springfield, Illinois.

Hendrickx, A.G. and Kriewaldt, F.H. (1967). Observations on a controlled breeding colony of baboons. In *The baboon in medical research*, Vol. 2 (ed. H. Vagteborg), pp. 69–83. University of Texas Press, Austin.

Hodgen, G.D. and Nieman, W.H. (1975). Application of the sub-human primate pregnancy test kit to pregnancy diagnosis in baboons. *Laboratory Animal Science*, **25**, 757–9.

Hummer, R.L. (1967). Preventive medicine practices in baboon colony management. In *The baboon in medical research*, Vol. 2 (ed. H. Vagteborg), pp. 51–68. University of Texas Press, Austin.

Hummer, R.L. (1970). Observations on the feeding of baboons. In *Feeding and nutrition of non-human primates* (ed. R.S. Harris), pp. 183–203. Academic Press, New York.

Johnsen, D.O. and Whitehair, L.A. (1986). Research facility breeding. In *Primates. The road to self-sustaining populations* (ed. K. Benirschke), pp. 499–511. Springer-Verlag, New York.

Jones, M.L. (1968). Longevity of primates in captivity. *International Zoo Yearbook*, **8**, 183–92.

Kalter, S.S. (1977). The baboon. In *Primate conservation* (ed. Prince Rainier II of Monaco and G.H. Bourne), pp. 385–418. Academic Press, New York.

Kinzey, W.G. (1971). Male reproductive system and spermatogenesis. In *Comparative reproduction of non-human primates* (ed. E.S.E. Hafez), pp. 85–114. Charles C. Thomas, Springfield, Illinois.

Lee, P.C., Thornback, J., and Bennett, E.L. (1988). *Threatened primates of Africa*. The IUCN red data book. IUCN, Cambridge.

Love, J.A. (1978). A note on the birth of a baboon (*Papio anubis*). *Folia Primatologica*, **29**, 303–6.

Mack, D. and Kafka, H. (1978). Breeding and rearing of woolly monkeys, *Lagothrix lagotricha*, at the National Zoological Park, Washington. *International Zoo Yearbook*, **18**, 117–23.

Martin, D.P. (1986). Reproduction and obstetrics. In *Zoo and wild animal medicine* (ed. M.E. Fowler), pp. 701–4. W.B. Saunders Co, Philadelphia.

Max Lang, C. (1971). Techniques of breeding and rearing of monkeys. In *Comparative reproduction of non-human primates* (ed. E.S.E. Hafez), pp. 455–72. Charles C. Thomas, Springfield, Illinois.

Michael, R.P. and Zumpe, D. (1971). Patterns of reproductive failure. In *Comparative reproduction of non-human primates* (ed. E.S.E. Hafez), pp. 205–42. Charles C. Thomas, Springfield, Illinois.

Miller, R.L. and Pallota, A.J. (1965). Comments on the maintenance of a small baboon colony. In *The baboon in medical research* (ed. H. Vagteborg), pp. 111–24. University of Texas Press, Austin.

Moore, G.T. and Cummins, L.B. (1979). Nursery rearing of infant baboons. In *Nursery care of non-human primates* (ed. G.C. Ruppenthal), pp. 145–51. Plenum Press, New York.

Napier, J.R. and Napier, P.H. (1967). *A handbook of living primates*. Academic Press, London.

Newsome, J. (1967). Baboons. In *The UFAW handbook on the care and management of laboratory animals* (3rd edn), pp. 709–19. Livingstone, Edinburgh & London.

Nowell, L.H., Heidrich, A.G., and Appolynaire, S. (1978). Labour and parturition in feral olive baboons Papio anubis. In *Recent advances in primatology*, Vol. 1 (ed. D.J. Chivers and J. Herbert), pp. 511–14. Academic Press, London.

Oftedal, O.T. (1984). Milk composition, milk yield, and energy output at peak lactation: a comparative review, *Symposia of the Zoological Society of London*, **51**, 33–85.

Ott-Joslin, J.E. (1986). Viral diseases in non-human primates. In *Zoo and wild animal medicine* (ed. M.E. Fowler), pp. 647–97. W.B. Saunders, Philadelphia.

Reed, O.M. (1965). Studies of the dentition and eruption patterns in the San Antonio baboon colony. In *The baboon in medical research* (ed. H. Vagteborg), pp. 167–89. University of Texas Press, Austin.

Richard, A.F. (1985). *Primates in nature*, pp. 207–29. W.H. Freeman, New York.

Rowell, T.E., Din, N.A., and Omar, A. (1968). The social development of baboons in their first three months. *Journal of Zoology, London*, **155**, 461–83.

Sigg, H., Stolba, A., Abegglen, J.J., and Dasser, V. (1982). Life history of Hamadryas baboons: physical development, infant mortality, reproductive parameters and family relationships. *Primates*, **23**, 473–87.

Smuts, B.B. (1985). *Sex and friendship in baboons*. Aldine, New York.

Snow, C.C. (1967). Some observations on the growth and development of the baboon. In *The baboon in medical research*, Vol. 2 (ed. H. Vagteborg), pp. 187–99. University of Texas Press, Austin.

Vice, T.E., Britton, H.A., Ratner, I.A., and Kalter, S.S. (1966). Care and raising of newborn baboons. *Laboratory Animal Science*, **16**, 12–22.

Whitney, R.A. and Wickings, E.J. (1986). Macaques and other old world simians. In *The UFAW handbook on care and management of laboratory animals* (ed. T.B. Poole), pp. 599–627. Longman, Harlow.

Proboscis monkey (© New York Zoological Society)

15 Proboscis monkey

Species
The proboscis monkey   *Nasalis larvatus*

ISIS No.   1406008007001001

Status, subspecies, and distribution
There are no subspecies according to the ISIS classification. The species is found in Borneo (Medway 1970; Hrdy 1977) and on two small islands close to its north-eastern coast: Berhala and Sebatik (Wolfheim 1983). It is a forest species most commonly found in riverine or coastal areas (Kawabe and Mano 1972; Happel *et al.* 1987).

The population appears to have been badly affected at even low cutting rates of the Bornean forests (Wilson and Wilson 1975) and its status is vulnerable (Marsh 1987, IUCN 1990). MacKinnon (1986) estimated the total wild population to be about 260 000.

The species is not used in biomedical research and the captive population is small. It was first bred in captivity at San Diego in 1965 (Jones 1986).

Sex ratio
No data are available on the sex ratio at birth.

Social structure
Proboscis monkeys live in groups of 10 to 32 individuals (Napier and Napier 1967; Kawabe and Mano 1972), with an average of about 20 (Happel *et al.* 1987). Troops of up to 40 to 50 individuals have been recorded (Kern 1964). Troops consist of several males with a larger number of mature females and juveniles (Kawabe and Mano 1972; Hrdy 1977). The average composition of a troop in one survey (Macdonald 1982) was 1:2.6:1.8 (males to females to juveniles). Groups may split temporarily for foraging (Macdonald 1982) but keep to their home ranges which are about 100 to 150 hectares.

Breeding age
No specific information is available.

Longevity
Insufficient data are available.

Seasonality
Data derived from 15 females shot in the wild did not indicate any seasonality in births (Schultz 1942; see also Struhsaker and Leland 1987).

Gestation
The gestation period is thought to be about 166 days (Happel *et al.* 1987).

Pregnancy diagnosis
No specific data are available.

Birth
One female showed a change in behaviour and became aggressive as she approached parturition. She gave birth to one young during the night and the placenta was probably eaten (Pournelle 1967). Hollihn (1973), however, reported that births occurred in the morning.

Litter size
The usual litter size is 1, but Schultz (1956) recorded one set of twins.

Adult weight
Adult males weigh about twice as much as females (Kern 1964). The weight of a males ranges from 14 to 23.6 kg with an average of 20.3 kg and the range in females is 8.2 to 11.8 kg with an average of 9.9 kg (Schultz 1942), or 10 kg (Happel *et al.* 1987).

Neonate weight
Schultz (1942) estimated birth weight at 0.45 kg or 4.6 per cent of maternal body weight. One live

male infant presented as a breech birth and was delivered by Caesarean section. It weighed 0.6 kg (Ruedi 1981).

Adult diet

The proboscis monkey feeds on the leaves and shoots of trees (Medway 1970), especially those of the mangrove tree but not the nipa-palm (Kawabe and Mano 1972). In one report, 95 per cent of the diet of this species was made up of the leaves of the mangrove tree (*Rhizophora apiculata*) and the pedada tree (*Sonneratia apiculata*) but it was thought that some fruits and flowers were also taken (Kern 1964).

Pournelle (1961) recommended that recently captured or imported proboscis monkeys should initially be fed foliage, such as leaves, stems, and berries of *Acacia* species, *Eugenia* species, *Solandra guttata*, and flowers of *Hibiscus rosasinensis*. Other species that are accepted are *Salix*, *Polygonum aubergii*, and *Robinia pseudoacacia* (Hollihn 1973). Later on, supplementary foods may be offered, such as lettuce, yams, sweetcorn, potatoes, celery, bread, peanuts, and a selection of fruits (Pournelle 1961). It appears that this species, like other leaf-eaters, prefers unripe fruit and bananas, citrus fruit, and melons can be provided unpeeled. Foods containing easily digestible carbohydrates and fats, such as bread, rice, nuts, oats, and yams, should be used sparingly and with caution as they may cause 'bloat' (distension of the stomach due to rapid fermentation, Hill 1964; Hollihn 1973). Although leaf-eating monkeys may adapt to foods other than their natural diet, and may prefer them, foliage should always form a part of their diet.

Wackernagel (1977) described the composition and preparation of a 'cake' based on cereals supplemented with other sources of energy — animal and vegetable protein, minerals and vitamins mixed with cooked minced meat. This cake was readily accepted by leaf-eating species. He also listed the species of trees whose branches and leaves could safely be given.

Adult energy requirements

As far as we are aware, there is no information available for this species.

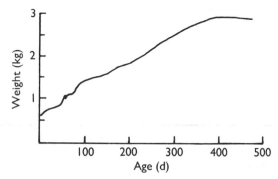

Fig. 15.1. Growth rate of a hand-reared male proboscis monkey. From Ruedi (1981).

Growth

The weight gain of one hand-reared male has been reported by Ruedi (1981). This infant suffered several setbacks and died at the age of 478 days and the growth curve may not be typical of the species (Fig. 15.1). During the first year the daily weight gain of this animal was about 6 g/d, which is similar to that of other Old World monkeys, such as the baboon, which have a comparable adult weight.

Milk and milk intake

There are no data on the composition of the milk of this species. One proboscis infant was reared on a human milk replacer (Ruedi 1981; see below). Human milk replacers have been used to rear other colobine monkeys (see western black and white colobus monkeys, Chapter 16), but Primilac may mimic the composition of the natural milk more closely. Hill (1964) mentioned that unmodified cow's milk was not appropriate for very young proboscis monkeys and that a human milk replacer was preferable.

The daily milk intake of Ruedi's (1981) hand-reared infant proboscis monkey is shown in Fig. 15.2. The volume taken each day increased from 80 ml on the first day to about 200 ml/d from 50 days of age. The daily energy intake, assuming the milk contained 0.67 kcal/ml suggests a relatively low rate of energy intake in relation to metabolic weight: from less than 100 in the first days to a maximum of about 150 kcal/d per $kg^{0.75}$.

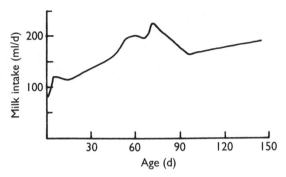

Fig. 15.2. Daily milk intake of a hand-reared proboscis monkey in relation to age. From Ruedi (1981).

Lactation and weaning

In captivity, infants first begin to test solids by picking left-over foods with their mouths when they start moving away from their mothers at 4 to 6 weeks of age (Pournelle 1967; Hollihn 1973). At 2 months of age they begin to use their hands to eat very small amounts of solids but regular ingestion of solids is not seen until 6 months of age (Hollihn 1973). At that age, a female proboscis began to resist suckling attempts by her infant (Pournelle 1967), and by 7 months of age this infant appeared to be completely weaned. It did occasionally suck after that stage but apparently for security rather than nutrition.

Feeding

There is one account of the hand-rearing of one infant proboscis monkey given by Ruedi (1981). The first feed given to the baby was camomile tea. The second feed was a mixture of camomile tea, a human milk replacer, and supplementary vitamins. The formula was given by bottle every 2 hours from days 1 to 12 and, thereafter until approximately 3 months of age, feeds were given every 2.5 hours. At 97 days of age, all night feeds were discontinued. From day 145, the infant was fed seven times daily and this was reduced to six times daily on day 285. The infant suckled well but overfeeding resulted in regurgitation and vomiting. A variety of other vitamin and mineral supplements were used during the hand-rearing process but the infant suffered numerous gastrointestinal upsets and setbacks and died at 478 days of age. Ruedi considered that the problems were primarily due to stress as the infant was separated from its social group and experienced a number of changes of environment.

The techniques used for hand-rearing macaques and other Old World monkeys (see, for example, rhesus macaque and baboon) are probably largely applicable to this species.

Accommodation

The baby hand-reared by Ruedi (1981) was initially housed in a wooden box (47 cm long, 30 cm wide and 37 cm high) warmed by a heating pad to 30 °C. It was provided with a fur-covered surrogate mother. The room temperature was 25 °C and the relative humidity 55 per cent. From one month of age, the infant was allowed to spend short periods outside every day. However, due to illness (possibly influenza), the infant was hospitalized in an incubator and then housed in an isolette in the nursery for four weeks. By the age of 79 days, the infant was allowed to exercise in a play-pen for 30 minutes daily and was then gradually introduced to the adult proboscis monkey enclosure two months later. By day 144 the infant was left overnight in an area partitioned-off within the adult's enclosure (Ruedi 1981).

Infant management notes

No specific data are available.

Physical development

At birth the face is bright blue (Pournelle 1966, 1967; Hrdy 1977) and the nose is small and uptilted (Pournelle 1967). At 2.5 to 3 months of age the face becomes a dark sooty grey (Pournelle 1966, 1967). By 7 months of age the neck 'ruff' is still poorly defined and the beard and cheek whiskers are brown and inconspicuous and do not yet form a sharp facial frame. The head cap is darker than the back but not as dark as in adults and the ears are exposed. The lumbar patch is also visible but not yet well defined and the shoulders and limbs are brown rather than the salt and pepper grey of adults. By 8.5 months of age, the face is still grey but lightens in colour gradually to the flesh-coloured face of adults. The nose in females remains small and slightly uptilted but in males it increases very gradually in size until it becomes pendulous when mature, or by 6 years of age (Pournelle 1967). The sequence of eruption of the deciduous teeth was described by Schultz (1942). The order is as follows: upper middle incisors, lower lateral incisors, lower middle and upper lateral incisors, lower and upper first molars, lower canines, upper canines, lower second molars, and, finally, upper second molars. The sequence of eruption of the permanent teeth (Schultz 1942) is: lower first molars, upper first molars, followed by a resting period, then the lower middle incisors, upper middle incisors, lower lateral incisors, upper lateral incisors, lower second molars, upper second molars, lower and upper first and second premolars (in an unknown but probably rapid and irregular succession). In females, the order of eruption continues as follows: lower canines, upper canines, lower third molars, and upper third molars, whereas in males, the lower third molars erupt before the simultaneous eruptions of the upper third molars and lower canines, and the upper canines appear last. Swindler (1976) observed that the eruption sequence appeared to be more like that of *Cercopithecus* species than other *Colobidae* leaf-eaters, as the second molars erupted after the middle incisors.

Behavioural development

Infants are always carried on the chest (Kern 1964). When infants initially test solid foods, they pick them up with their mouths. At 2 months of age they begin to use their hands and by 3 months of age they use their hands exclusively for picking up objects or food. During the first year of life, Colobids bite leaves off twigs but later pick single leaves with one hand while gripping the branch with the other. This dexterity improves so rapidly that by 2 years of age they can strip the leaves off a branch with one movement (Hollihn 1973). Male proboscis monkeys begin to show sexual behaviour patterns by 1 year of age (Hollihn 1973).

Disease and mortality

There is very little specific information. The infant reared by Ruedi (1981) eventually died after several bouts of illness including episodes of vomiting and diarrhoea, fever, and laboured breathing. Some of these problems were thought to be associated with overfeeding on occasions. At postmortem there was evidence of alimentary tract candidiasis, but this may have been a terminal infection. Ruedi also reported the death of one female and her infants from nephrolithiasis, which suggested a genetic predisposition (although environmental factors could explain this). Hill (1964) suggested that the poor longevity records for *Colobidae* in captivity at that time were due to mortality, in at least half the cases from digestive upsets caused by inappropriate diet, in particular gastric dilatation or 'bloat'.

Preventative medicine

There are no specific data for this species.

Indications for hand-rearing

No specific data are available. It is possible that, as in colobus monkeys, females other than the mother may carry the infant, so careful observation would be necessary before concluding that an infant had been 'stolen' as sometimes occurs in macaques (Hill 1972).

Reintegration

The baby hand-reared by Ruedi (1981) was first introduced to the empty adult proboscis monkey enclosure over a period of time before the adults were returned. A keeper remained with the infant and supervised the interactions between the adults and the youngster. These interactions appeared positive and complete reintroduction was viewed with optimism and may have been successful had the infant survived.

References

Happel, R.E., Noss, J.F., and Marsh, C.W. (1987). Distribution, abundance and endangerment of primates. In *Primate conservation in the tropical rain forest* (ed. C.W. Marsh and R.A. Mittermeier.), pp. 63–82. Alan R. Liss, New York.

Hill, C.A. (1972). Infant sharing in the family Colobidae emphasizing *Pygathrix*. *Primates*, **13**, 195–200.

Hill, W.C.O. (1964). The maintenance of langurs (Colobidae) in captivity; experiences and some suggestions. *Folia Primatological*, **2**, 222–31.

Hollihn, U. (1973). Remarks on the breeding and maintenance of Colobus monkeys *Colobus guereza*, Proboscis monkeys *Nasalis larvatus* and Douc langurs *Pygathrix nemaeus* in zoos. *International Zoo Yearbook*, **13**, 185–8.

Hrdy, S.B. (1977). *The langurs of abu: female and male strategies of reproduction*. Harvard University Press, Cambridge, Mass.

IUCN (1990). *1990 IUCN red list of threatened animals*. pp. 13. IUCN, Gland, Switzerland.

Jones, M.L. (1986). Successes and failures of captive breeding. In *Primates. The road to self-sustaining populations* (ed. K. Benirschke), pp. 251–60. Springer-Verlag, New York.

Kawabe, M. and Mano, T. (1972). Ecology and behaviour of the wild proboscis monkey *Nasalis larvatus* (Wurmb) in Sabah, Malaysia. *Primates*, **13**, 213–28.

Kern, J.A. (1964). Observations on the habits of the proboscis monkey *Nasalis larvatus* WURMB, made in the Brunei Bay area, Borneo. *Zoologica (New York)*, **49**, 183–92.

Macdonald, D.W. (1982). Notes on the size and composition of groups of proboscis monkey *Nasalis larvatus*. *Folia Primatologica*, **37**, 95–8.

MacKinnon, K. (1986). The conservation status of nonhuman primates in Indonesia. In *Primates. The road to self-sustaining populations* (ed. K. Benirschke), pp. 99–126. Springer-Verlag, New York.

Marsh, C.W. (1987). A framework for primate conservation priorities in Asian moist tropical forests. In *Primate conservation in the tropical rain forest* Alan R. Liss, New York. (ed. C.W. Marsh and R.A. Mittermeier), p. 343–54.

Medway, Lord (1970). The monkeys of Sundaland. In *Old world monkeys, evolution, systematics and behaviour*. Academic Press, London. (ed. J.R. Napier and P.H. Napier), pp. 513–53.

Napier, J.R. and Napier, P.H. (1967). *A handbook of living primates*. Academic Press, London.

Pournelle, G.H. (1961) Observations on captive proboscis monkeys (*Nasalis larvatus*). *International Zoo Yearbook*, **3**, 69–70.

Pournelle, G.H. (1966). Birth of a proboscis monkey. *Zoonooz* (Newsletter of the Zoological Society of San Diego) **39**, 3–7.

Pournelle, G.H. (1967). Observations on reproductive behaviour and early postnatal development of the proboscis monkey (*Nasalis larvatus orientalis*). *International Zoo Yearbook*, **7**, 90–2.

Ruedi, D. (1981). Hand-rearing and reintegration of a caesarian-born proboscis monkey, *Nasalis larvatus* at Basle Zoo. *International Zoo Yearbook*, **21**, 225–9.

Schultz, A.H. (1942). Growth and development of the proboscis monkey. *Bulletin of the Museum of Comparative Biology at Harvard University*, **89**, 274–314.

Schultz, A.H. (1956). The occurrence and frequency of pathological and teratological conditions and of twinning among non-human primates. In *Primatologia*, Vol. 1 (ed H. Hofer, A.H. Schultz, and D. Stark), pp. 965–1014. Karger, Basel.

Struhsaker, T.T. and Leland, L. (1987). Colobines: infanticide by adult males. In *Primate societies* (ed. B.B. Smuts, R.M. Cheney, R.W Seyfarth, T.T. Wrangham, and T.T. Struhsaker), pp. 83–97. University of Chicago Press.

Swindler, D.R. (1976). *Dentition of living primates*, pp. 142–5. Academic Press, London.

Wackernagel, H. (1977). Feeding of apes and monkeys at Basle Zoo. *International Zoo Yearbook*, **171**, 189–94.

Wilson, C.C. and Wilson, W.L. (1975). The influence of selective logging on primates and some other animals in East Kalimantan. *Folia Primatologica*, **23**, 245–74.

Wolfheim, J.H. (1983). *Primates of the world*, pp. 533–6. University of Washington Press, Seattle.

Western black and white colobus monkey

16 Western black and white colobus monkey

Species
The western black and white colobus monkey *Colobus polykomos*

ISIS No.   1406008006004001

Status, subspecies, and distribution
Some authorities consider this species, which occurs in the countries of the southern coast of West Africa from Senegal to Nigeria, to be distinct from *C. angolensis* which occurs in central Africa (Wolfheim 1983). Others have classified these as one species (Napier and Napier 1967). ISIS includes *angolensis* with *polykomos* and recognizes 12 subspecies. The western black and white colobus has been described as common in some parts of Ghana and the Ivory Coast but it is very rare or extinct in Togo and Nigeria (Wolfheim 1983). These forest animals are found in a wide range of forest types (Wolfheim 1983). They are largely arboreal, only descending to the ground for salt or other minerals (Ormrod 1967).

The black and white colobus was seriously threatened by the European fur trade at the end of the nineteenth century (Brandon-Jones 1984) and it is still hunted for fur and meat (Booth 1979; Wolfheim 1983; Oates *et al.* 1987). Habitat destruction is also threatening the status of *C. polykomos* in parts of its range (Struhsaker and Leland 1987). Lee *et al.* (1988) classified the species as not threatened, and it is not included in the IUCN red list.

The species has not been used for biomedical research as far as we aware, and is not widely kept in zoos. Fifteen *C. polykomos* and six unidentified *Colobus* species were imported into the United States between 1968 and 1973 (Wolfheim 1983), and 150 *C. polykomos* were imported into the United Kingdom between 1965 and 1975 (Burton 1978). Olney (1988) recorded six births in captivity during 1985.

Sex ratio
The sex ratio at birth is probably close to 1:1. Among the 11 births recorded by Coffey (1970) there were 5 males and 6 females. The records of captive births between 1980 and 1985 listed in the *International Zoo Yearbook* (Olney 1988) show that of 37 babies sexed, 17 were male and 20 were female.

Social structure
Emerson (1973) observed social groups in the wild consisting of approximately eight animals, including one or occasionally two adult males, three to four females, one to two subadults, and one infant. Ormrod (1967) observed groups of 12 animals whereas Coffey (1970) recorded family groups or larger units up to 20 animals. Dominance relationships do not appear to be very strict (Ormrod 1967).

Breeding age
Both sexes reach sexual maturity at about 4 years of age (Brandon-Jones 1984).

Longevity
One specimen of *C.p. kikuyensis* was reported to have lived in captivity for 23.5 years, and a *C.p. angolensis* lived for 16.5 years (Jones 1968).

Seasonality
Colobines in the wild do not appear to adhere to a distinct breeding season (Struhsaker and Leland 1987), although there does seem to be some temporal clustering of births which may possibly coincide with an abundance of foods at weaning time (Brandon-Jones 1984). In captivity, Coffey

(1970) recorded no seasonal pattern among 10 births, but there were none in December, January, and February. *Colobus polykomos* is said to produce one offspring yearly (Hill 1972), and one female who reared her infant in captivity produced another after 380 days. The interval may be shorter following a still birth or neonatal death (for example, 216 days, Coffey 1970).

Gestation

The gestation periods for five pregnancies were 147, 154, 169, 177, and 178 days, with an average of 165 days (Mallinson 1972). Previously, Coffey (1970) had estimated 178 days.

Pregnancy diagnosis

No specific tests have been described for this species. It has been noted that abdominal distension is difficult to see in this species (Coffey 1970).

Birth

Births usually occur at night and one delivery has been described by Coffey (1970). Straining was irregular and the female attempted to pull on the infant's head. When one shoulder appeared, the mother lay on her side with her eyes closed, still pulling at the infant. Strong contractions then followed and a large stillborn infant was delivered. The placenta was probably eaten.

Litter size

The litter size is usually 1, but twins have been reported (Coffey 1970).

Adult weight

Males are heavier than females. One adult male weighed 12.5 kg and three adult females weighed 9.5, 9.0, and 8.5 kg (Mallinson 1969). Two other males captured from the wild weighed 9.5 and 12.3 kg (Ohwaki *et al.* 1974).

Neonate weight

Babies weigh about 400 g at birth (Brandon-Jones 1984).

Adult diet

In the wild, *Colobus* monkeys feed on leaves (Bauchop and Martucci 1968; Hollihn 1973), mosses, and other coarse green matter (Ormrod 1967), and fruit and seeds (Kay *et al.* 1976; Struhsaker and Leland 1987). The stomach contents of two wild-caught *C. polykomos* specimens in Kenya consisted primarily of starchy seeds and fruit with some leaves (Ohwaki *et al.* 1974), whereas those of *C. polykomos* (in Ghana) contained no fruit or animal matter (Booth 1979).

Foliage is considered to be an essential part of the diet of *Colobus* monkeys in captivity (Hill 1964; Ormrod 1967; Mallinson 1969) and the leaves and branches of beech, lime, elm, hawthorn (Mallinson 1969), willow, and *Robinia* (Klos 1966) have been used. The leaves of *Polygonum aubergii*, *Robinia pseudoacacia*, and *Salix* species are particularly favoured (Hollihn 1973). Bamboo and the buds and bark of willow branches have been offered during winter to colobines in captivity in Europe (Mallinson 1969). Foliage collected during the spring and summer has been frozen for feeding during the winter (Hollihn 1973). Wackernagel (1977) compiled a list of European trees that appear safe to feed to *Colobus* monkeys, and he and Klos (1966) described the preparation and ingredients of a 'cake' based on cereals with additional sources of protein and energy which is readily accepted by this species. Recently, proprietary high-fibre diets have been designed for leaf-eaters (Watkins *et al.* 1985). Diets may also be supplemented by a limited amount of fresh fruits and vegetables (Klos 1966; Ormrod 1967; Mallinson 1969; Hollihn 1973; Wackernagel 1977) although care should be taken to avoid overloading the animal's digestive capacity with large amount of carbohydrates which may result in 'bloat' (Hill 1964; Hollihn 1973).

Adult energy requirements

There are no specific data available. Measurements of food intake at the Zoological Society of London indicated a daily energy intake of about 140 kcal per metabolic weight.

Growth

Colobus monkeys reach their full adult size at approximately 5 years of age (Brandon-Jones 1984). The growth of young *C. polykomos* has not been described as far as we are aware. The weight

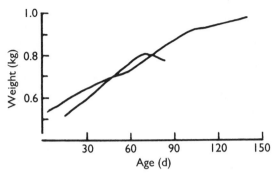

Fig. 16.1. Growth rate of two hand-reared *Colobus guezera*. From Anon (1985) and Velte (1985).

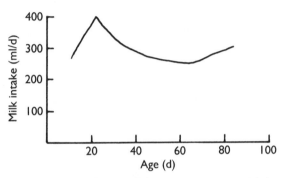

Fig. 16.2. Daily milk intake of a hand-reared infant black and white colobus monkey *C. polykomos*. From Usher-Smith (1972).

gain of two hand-reared infant *Colobus guezera* (anon. 1985; Velte 1985) is shown in Fig. 16.1.

Milk and milk intake

There are no data on the composition of the milk of this species. Human milk replacers, for example, SMA, Similac, and Enfamil, have been used to hand-rear colobus babies (Usher-Smith 1972; anon. 1985; Velte 1985). The milk intake of the baby reared by Usher-Smith (1972) is shown in relation to age in Fig. 16.2.

Lactation and weaning

Infants appear to remain in contact with their mothers nipples for long periods without actually feeding. One infant was observed tasting hawthorn leaves at 6 days of age (Mallinson 1969), Usher-Smith (1972) first offered solid foods at 19 days of age and within a week the baby was regularly eating Farex, chopped leaves, and peeled grapes. Velte (1985) began weaning a hand-reared *C. guezera* at 10 weeks and by 13 weeks the baby was no longer receiving any milk, and baby cereal had become 'the major part of the diet, with soaked monkey chow and chopped fruit also being provided. Weaning of a baby *C. guezera* hand-reared at St. Louis Zoo began by adding ground monkey chow to the milk at day 43 and by offering chopped fruit and vegetables from day 48.

Feeding

One 11-day-old infant *C. polykomos* was removed for hand-rearing (Usher-Smith 1972). The formula used was SMA mixed with water. The initial ratio of SMA scoops to fluid ounces of water was 3:3 (i.e., 15 g SMA powder to 85 ml). However, when the infant developed diarrhoea, the formula was further diluted and so the ratios varied with each feed. By the age of 22 days, the ratio was increased to 3:4 and gradually reduced to 3:3.5 by day 32 onwards. The milk intake per feed was restricted to 45 ml initially. On day 22, this was further restricted to 40 ml per feed. Between the ages of 26 and 64 days, the infant was offered 45 ml per feed. On day 64, *ad libitum* milk was offered in a self-feeding bottle. The average daily intake of this infant is shown in Fig. 16.2 (Usher-Smith 1972).

Usher-Smith (1972) initially used a human baby feeder bottle fitted with a small teat designed for kittens for hand-feeding. This was gradually replaced by a laboratory-type glass drinking bottle with a stainless steel nipple on aluminium tubing for self-feeding. The baby bottle was sterilized between feeds and the milk was warmed to 35 °C. The self-feeding glass bottle was also sterilized between each feed. The infant was initially fed every 2 hours, providing a total of 12 feeds per 24 hours. By day 22, the infant was offered a total of 10 feeds every 2 hours except at night. By day 26, a total of 8 feeds were offered every 2.5 hours, which was reduced to 7 feeds every 3 hours on day 32 and reduced further to a total of 6 feeds every 3.5 hours on day 44. On day 64, a self-feeding bottle containing 70 ml of formula was provided, although the hand-feeding schedule was continued. The bottle was replenished at feeding times. By day 84, all feeds were obtained *ad libitum* from the self-feeding bottle and hand-feeding ceased.

A baby *C. guezera* hand-reared at St. Louis Zoo was fed only 5 times daily from birth (anon. 1985).

Usher-Smith (1972) first offered solids, in the form of fresh twigs and leaves, when the baby was observed chewing on cotton-wool at the age of 19 days. By day 25, Farex baby cereal mixed with condensed milk, chopped leaves, and peeled grapes were all being offered and taken regularly. At 44 days of age, the infant was offered three daily feeds of Farex cereal, chopped leaves, diced mixed fruits, and fresh twigs of hawthorn, sweet chestnut, elm, oak or flowering currant in addition to the milk formula. A dish of water was provided on day 84 when hand-feeding ceased.

Accommodation

One infant separated from its mother at the age of 11 days was housed in a glass-fronted hospital cage (Usher-Smith 1972). The floor was lined with cotton-wool and the walls with cloth padding. The temperature was maintained at 30 °C. The infant's cage was provided with two horizontal branches. At St. Louis Zoo a baby was kept in an incubator for the first 16 days then moved to a small cage with a heat lamp.

Infant management notes

A soft surrogate should be provided for the young infant to cling to. See the section on the squirrel monkey. (Chapter 10.)

Physical development

The eyes are open at birth, and the coat is short and markedly different in colour to that of adults, being nearly all white and grey (Ormrod 1967; Horwich and Manski 1975), and there is less skin pigment (Brandon-Jones 1984). Although the coat begins to darken within a few days, the adult colour is not attained until 120 days of age or more (Ormrod 1967; Horwich and Manski 1975; Mearns and Pidgeon 1978). Infants may (Usher-Smith 1972) or may not (Ormrod, 1967) have nails on their rudimentary thumbs. The deciduous incisors are the first teeth to appear and one infant was born with four lower teeth already erupted (Horwich and Manski 1975). In another infant of 11 days of age, all upper and lower incisors were present (Usher-Smith 1972). The sequence of eruption of teeth appears to be similar to other Old World monkeys except that in the permanent dentition, the second upper and lower molars have a tendency to erupt early, before the second incisors (Swindler 1976).

Mearns and Pidgeon (1978) described four stages of pelage development as observed for five infants. Although individual variation is considerable, particularly in the colouring of the tail, hands, and feet even at birth, and the timing and order of pelage changes, the general pattern may provide an indication of development (Mearns and Pidgeon 1978). From birth onwards, the neonatal coat colour darkens so that by the end of the first stage at 3–4 weeks of age, the tail has darkened considerably, the hands and feet and the ischial callosities have become black/grey, the face and dorsum light grey, the head cap grey, whereas the ventrum, the rest of the limbs, and the genitals are still white (Horwich and Manski 1975; Mearns and Pidgeon 1978). By the end of the second stage, at 6–7 weeks, the head cap, limbs, and ischial callosities are black, the body is becoming grey all over, as well as the face, and the genitals and tail appear light grey (Mearns and Pidgeon 1978). The hair over the shoulders is becoming long and more long white hairs appear at the tip of the tail (Horwich and Manski 1975). By the end of the third stage, at 9–11 weeks, the body is black, some white crest hairs are present in front of the black head cap (Horwich and Manski 1975; Mearns and Pidgeon 1978), the limbs appear like those of adults (Usher-Smith 1972; Mearns and Pidgeon 1978), the ischial callosities are black, the tail light grey with a white tip, and the genitalia grey (Mearns and Pidgeon 1978). When the end of the fourth stage is reached at 15–18 weeks, the young *C. polykomos* looks like an adult in its pelage pattern, with long white crest hairs, in front of the black head cap, a white 'cape' over the shoulders, and an otherwise black body, black limbs except for the upper arms, white tail, black genitals, and ischial callosities (Usher-Smith 1972; Mearns and Pidgeon 1978). In some infants, there appears to be a delay in the darkening of the face and it is thought this may be due to a lack of sunlight

(Usher-Smith 1972). Summer-born infants develop black faces at a much earlier age than those born in the winter (Mearns and Pidgeon 1978).

Behavioural development

The newborn *C. polykomos* is very inactive, clings to the ventrum of a female and can support its own weight in this position (Emerson 1973; Brandon-Jones 1984). Maternal care, apart from nursing, is mostly administered by females other than the mother during the infant's early life. For the first few weeks the infant is not very mobile and has little control over who carries it but by 3 weeks it increasingly recognizes its mother and starts returning to her (Horwich and Manski 1975) as at that stage it begins to move independently (Emerson 1973) and may also be seen sitting alone (Mallinson 1969). By 5 weeks of age its locomotor abilities have increased and it can resist other females to remain with its mother where it sleeps on her ventrum with a nipple in the mouth (Horwich and Manski 1975).

In one infant, play was first observed in the field at the age of 13 days (Mallinson 1969) and after three weeks some time is also spent alone (Horwich and Manski 1975). A hand-reared infant began using the horizontal branches in its cage at 29 days of age (Usher-Smith 1972). It was also observed sitting on perches in the adult stance on day 37. Play and self-grooming first appeared on day 42. On day 51, it showed the submissive mouth opening behaviour. By day 62, it was playing and mock-fighting and its jumping abilities developed so that it could jump a distance of 35 cm on day 78. On day 83, 'whoofing' noises were made during mock fights. By days 95 and 96, it began exploring further afield and jumping more than a metre (Usher-Smith 1972).

Disease and mortality

Neonatal or infant mortality in captivity has been relatively high in the past. In one report of 11 births (Coffey 1970), two were still births, two infants died within the first 30 hours, one at 152 days, one at 468 days, and five were still alive. In another report (Klos 1966) of eight births, two were premature, two were stillborn, one was rejected by its mother and died at the age of 5 days, and three were reared.

Among adults, gastrointestinal and/or dietary problems, and in particular 'bloat' have been a major cause of mortality in captivity in the past, particularly among newly-imported animals.

Preventative medicine

Fatal measles epidemics have been reported in recently imported groups of *Colobus*, so vaccination should be considered.

Indications for hand-rearing

Infants have been removed for hand-rearing because of poor mothering; including rejection and preventing suckling (Usher-Smith 1972; anon. 1985). In another report, a baby was taken because the mother died after parturition. It should be noted that among *C. polykomos* it is normal practice for other females to carry the infant (Hill 1972) and this should not be confused with baby-stealing that is sometimes observed in macaques.

Reintegration

Velte (1985) reported that there were no problems introducing a 7-month-old hand-reared infant to a group of three females and one male, and that the females were very protective and would pick up the baby when keepers approached. This 'aunting' behaviour may be a particular advantage for reintroduction of hand-reared *Colobus* to established groups, and could perhaps be exploited to provide socialization even before weaning.

References

Anon. (1985). Colobus monkey. In *Infant diet/care notebook* (ed. S.H. Taylor and A.D. Bietz). American Association of Zoo Parks and Aquariums, Wheeling, Virginia.

Bauchop, T. and Martucci, R.W. (1968). Ruminant-like digestion of the langur monkey. *Science*, **161**, 698–9.

Booth, A.H. (1979). The distribution of primates in the Gold Coast. In *Primate ecology: problem-oriented field studies* (ed. R.W. Sussman), pp. 139–53. Wiley, New York.

Brandon-Jones, D. (1984). Colobus and leaf monkeys. In *The encyclopaedia of mammals*, Vol. 1 (ed. D. MacDonald), pp. 398–405. Allen & Unwin, London and Sydney.

Burton, J.A. (1978). Primate imports in to the U.K. 1965–1975. In *Recent advances in primatology*, Vol. 2 (ed. D.J. Chivers and W. Lane-Petter), pp. 137–45. Academic Press, London.

Coffey, P.F. (1970). A breeding analysis of a group of captive black and white colobus, *Colobus polykomos polykomos*, Zimmerman, 1780, at the Jersey Wildlife Preservation Trust. *Report of the Jersey Wildlife Preservation Trust*, **7**, 10–15.

Emerson, S.B. (1973). Observations on infant sharing in captive *Colobus polykomos*. *Primates*, **14**, 93–100.

Hill, C.A. (1972). Infant sharing in the family *Colobidae* emphasizing *Pygathrix*. *Primates*, **13**, 195–200.

Hill, W.C.O. (1964). The maintenance of langurs (*Colobidae*) in captivity; experiences and some suggestions. *Folia Primatologica*, **2**, 222–31.

Hollihn, U. (1973). Remarks on the breeding and maintenance of colobus monkeys *Colobus guereza*, Proboscis monkey, *Nasalis larvatus* and Douc langurs *Pygathrix nemaeus* in zoos. *International Zoo Yearbook*, **13**, 185–8.

Horwhich, R.H. and Manski, D. (1975). Maternal care and infant transfer in two species of colobus monkeys. *Primates*, **16**, 49–73.

Jones, M.L. (1968). Longevity of primates in captivity. *International Zoo Yearbook*, **8**, 183–92.

Kay, R.N.B., Hoppe, P., and Maloiy, G.M.O. (1976). Fermentative digestion of food in the colobus monkey, *Colobus polykomos*. *Separatum Experentia*, **32**, 485–6.

Klos, H.G. (1966). A note on breeding the white-tailed colobus monkey, *Colobus polykomos caudatus*, at West Berlin Zoo. *International Zoo Yearbook*, **6**, 146.

Lee, P.C, Thornback, J., and Bennett, E.L. (1988). *Threatened primates of Africa*. The IUCN red data book. IUCN, Gland, Switzerland.

Mallinson, J.J.C. (1969). Observations on a breeding group of black and white colobus monkeys, *Colobus polykomos*. *International Zoo Yearbook*, **9**, 79–81.

Mallinson, J.J.C. (1972). Establishing mammals gestation periods at the Jersey Zoological Park. *Report of the Jersey Wildlife Preservation Trust*, **9**, 62–5.

Mearns, C.S. and Pidgeon, M.S. (1978). Breeding and pelage development of black and white colobus monkey *Colobus polykomos polykomos* (Zimmerman, 1780) at the Jersey Zoological Park. *Dodo*, **15**, 61–9.

Napier, J.R. and Napier, P.H. (1967). *A handbook of living primates*. Academic Press, New York.

Oates, J.F., Gartlan, J.S., and Struhsaker, T.T. (1987). A framework for African rain forest primate conservation. In *Primate conservation in the tropical rain forest* (ed. C.W. Marsh and R.A. Mittermeier), pp. 321–7. Alan R. Liss, New York.

Ohwaki, K., Hungate, R.E., Lotter, L., Hofmann, R.R., and Maloiy, G. (1974). Stomach fermentation in East African colobus monkeys in their natural state. *Applied Microbiology*, **27**, 713–23.

Olney, P.J. (ed.) (1988). *International Zoo Yearbook*, **27**, 407.

Ormrod, S. (1967). Ursine or black colobus. *Report of the Jersey Wildlife Preservation Trust*, **4**, 6–9.

Struhsaker, T.T. and Leland, L. (1987). Colobines: infanticide by adult males. In *Primate societies* (ed. D.L. Smuts, D.L. Cheney, R.M. Seyfarth, R.W. Wrangham, and T.T. Struhsaker), pp. 83–97. University of Chicago Press.

Swindler, D.R. (1976). *Dentition of living primates*. Academic Press, London.

Usher-Smith, J.H. (1972). Notes on the hand-rearing of an ursine or black and white colobus monkey, *Colobus polykomos*. *Report of the Jersey Wildlife Preservation Trust*, **9**, 26–31.

Wackernagel, H. (1977). Feeding of apes and monkeys at Basle Zoo. *International Zoo Yearbook*, **17**, 189–94.

Watkins, B.E., Ullrey, D.E. & Whetter, P.A. (1985). Digestibility of a high-fiber biscuit-based diet by black and white colobus (*Colobus guereza*). *American Journal of Primatology*, **9**, 137–44.

Wolfheim, J.H. (1983). *Primates of the world*. University of Washington Press, Seattle and London.

Velte, F.F. (1985). Colobus monkey. In *Infant diet/care notebook* (ed. S.H. Taylor and A.D. Bietz). American Association of Zoo Parks and Aquariums Wheeling, Virginia.

Lar gibbon

17 Lar gibbon

Species

The lar gibbon *Hylobates lar*

ISIS No.   1406009001005001

Status, subspecies, and distribution

Opinions differ about the taxonomy of gibbons, but Wolfheim (1983) recognized eight subspecies of *H. lar*: *agilis*, *moloch*, *muelleri*, *abbottii*, *vestitus*, *entelloides*, *carpenteri*, and *lar*. The species occurs from southern China, through Burma, Thailand, Malaysia, and Indonesia (Java, Sumatra, and Borneo) (Robbins Leighton 1987). It is highly arboreal and is found in many different types of forest (Wolfheim 1983).

Hylobates lar is threatened in the wild due to loss of its habitat by logging and cultivation (Wolfheim 1983; Robbins Leighton, 1987). The population in Indonesia has been estimated at 154 000 (Mackinnon 1986). The lar gibbon is listed in Appendix 1 of the Convention on International Trade in Endangered Species, but is not presently included in the IUCN red list.

The number in captivity in ISIS-registered establishments was 228 in 1985 (Flesness 1986) of which 127 had been born in captivity. The total captive population is probably considerably greater than this. The species is not widely used in biomedical research, but has been bred under laboratory conditions (Breznock *et al*. 1977).

Sex ratio

The sex ratio among adults in the field is 1:1 (Carpenter 1963; Chance and Jolly 1970). Among 49 births, 19 infants were male, 19 female, and 11 of unknown sex (Arnold 1973).

Social structure

Gibbons are monogamous and remain together for life (Carpenter 1963; Mackinnon 1977; Gittins and Raemaekers 1980; Robbins Leighton 1987). In the wild, *H. lar* is found in family groups of two to six animals consisting of one adult pair and up to four offspring of differing ages (Carpenter 1963; Ellefson 1974; Gittins and Raemaekers 1980; Robbins Leighton 1987). Groups of up to 15 animals have been recorded but these were thought to be the result of temporary encounters between two or more family units (Fooden 1971). Average group sizes of 3.5 (MacKinnon and MacKinnon 1978; Robbins Leighton 1987) to 4 (Chance and Jolly 1970; Gittins and Raemaekers 1980) animals have been recorded. The young are expelled from the family group when mature (Chance and Jolly 1970; Gittins and Raemaekers 1980; McKenna 1982), although they remain on the periphery of parental territories (Chance and Jolly 1970; Mackinnon and Mackinnon 1977; Gittins and Raemaekers 1980). Occasional solitary males are thus observed (Carpenter 1963; Chance and Jolly 1970) calling or singing to attract a female (Mackinnon and Mackinnon 1977; Gittins and Raemaekers 1980). A period of courtship precedes pairing, mating, and the formation of a new family unit (Mackinnon and Mackinnon 1977; Gittins and Raemaekers 1980). *Hylobates lar* is a territorial primate with a fixed home range (Carpenter 1963; Robbins Leighton 1987) and no sexual difference in dominance is observed (Carpenter 1963).

Breeding age

Sexual maturity is reached by both males and females at 7 to 8 years of age (Keeling and McClure 1972; Zuckerman 1981; Robbins Leighton 1987). One captive female came into oestrus at approximately 4 years of age once only and then resumed oestrous cycles again at approximately 6 years of age (Brody and Brody 1974). The reproductive lifespan of the female is 10 to 20 years (Carpenter 1963; Robbins Leighton 1987) and a female may produce on average five or six offspring (Robbins Leighton 1987).

Longevity

Hylobates lar may live for 30 years or more (Carpenter 1963; Jones 1968; Keeling and McClure 1972; Gittins and Raemaekers, 1980).

Seasonality

There is no clear seasonal distribution of births in the wild (Robbins Leighton 1987) or in captivity (Keeling and McClure 1972; Arnold 1973), although a peak in births around the turn of the year has been observed (Chivers and Raemaekers 1980). This may be due to synchronization of either mating or weaning with fruiting and flowering periods in the field (Chivers and Raemaekers 1980; Robbins Leighton 1987). Intervals between live births are about two to three years (Carpenter 1963; Ellefson 1974; Gittins and Raemaekers 1980; McKenna 1982). The menstrual cycle is 27 to 29 days in duration, and menstrual flow occurs over approximately three days (Badham 1967; Keeling and McClure 1972). No distinctive sexual swelling occurs in this species (Napier and Napier 1967) but a premenstrual swelling of the vaginal area (Badham 1967), with a change in colour and turgidity of the labia, is observed (Keeling and McClure 1972) when the female is most receptive. One female began oestrous cycles within one month when her 5.5-month-old infant was removed (Badham 1967).

Gestation

The gestation period in *H. lar* is 190 to 214 days (Breznock *et al.* 1979) or, on average, 210 days (Keeling and McClure 1972; Brody and Brody 1974; Martin *et al.* 1979).

Pregnancy diagnosis

Pregnancy can be detected by an increase in abdominal size (Berkson and Chaicumpa 1969) at three months of gestation (Badham 1967). Mammary gland enlargement may also be noticed at this stage (Badham 1967) or one month before parturition (Brody and Brody 1974). After conception, the absence of cycling and vulval swelling may aid in pregnancy diagnosis; these reappear three months post-partum (Berkson and Chaicumpa 1969). A commercial human pregnancy test was used to diagnose pregnancy at 2.5 months gestation (Brody and Brody 1974).

Pregnancy could probably be detected at about one month by ultrasonography, and later by abdominal palpation.

Birth

Birth occurs at night (Crandall 1945; Sasaki 1962; Badham 1967). Imminent parturition is signalled by a widening of the separation of the two ischial callosities (Brody and Brody 1974). At two births in captivity, the placenta was partially eaten but the umbilical cord was not severed by the mother, thus necessitating human intervention (Brody and Brody 1974). In one account of an evening birth the amniotic sac was seen to be protruding. The female pulled at it before the infant was delivered within one minute. The female then held the infant and cleaned it (Brody and Brody, 1974).

Litter size

The usual litter size is 1, but a wild female has been observed carrying twins, 9–12 months old (Ellefson 1974).

Adult weight

Adults weigh about 5 to 6 kg (Mackinnon and Mackinnon 1978; Gittins and Raemaekers 1980; Robbins Leighton 1987). There is little sexual dimorphism (Schultz 1973). Records for males range between 4.1 and 7.9 kg and for females between 3.9 and 6.8 kg (Napier and Napier 1967; Fooden 1971; Schultz 1973).

Neonate weight

Neonates are thought to weigh approximately 400 g at birth (Sasaki 1962; Rumbaugh 1966; Keeling and McClure 1972; Schultz 1973). The birth weights of the six babies reared by (Martin *et al.* 1979) ranged from 264 g (a premature but live birth) to 437 g, with a mean of 370 g. Excluding the premature baby, the mean birth weight was 391 g. Breznock *et al.* (1979) reported higher birth weights in the range 336 to 505 g.

Adult diet

Gibbons are both frugivorous and herbivorous (Carpenter 1963). Their preferred food in the wild are fruits and flowers (Mackinnon 1977; Gittins and Raemaekers 1980) and, in particular, ripe sugar-rich juicy fruits (Robbins Leighton 1987) that provide easily accessible energy (Raemaekers 1979). At least 60 different species of fruit are taken (Ellefson 1974) but figs form a major part of their diet (Mackinnon and Mackinnon 1978; Gittins and Raemaekers 1980). Fruits may form two thirds of the diet according to the season (Ellefson 1974; Mackinnon and Mackinnon 1978; Robbins Leighton 1987). Young leaves and shoots are also taken and may account for a third of the diet (Ellefson 1974; Aldrich-Blake 1978; Robbins Leighton 1987). Insects, such as spiders, ants, termites, caterpillars, and stick insects, are also taken (Ellefson 1974; Gittins and Raemakers 1980) and may account for 4 per cent of the diet (Mackinnon and Mackinnon 1978). Eggs and nestlings may also be taken (Carpenter 1963). Gibbons spend most of the day foraging (Robbins Leighton 1987) and are very selective feeders (Raemaekers 1978a; Gittins and Raemaekers 1980). Fruits and especially figs are taken in the first and last feeds of the day, whereas leaves (for example, *Sloetia* and *Diospyros*, Raemaekers 1978b) are taken in the afternoon (Raemaekers 1978a).

In captivity, gibbons show a preference for fresh fruit and vegetables but do accept commercial monkey diets (Badham 1967; Keeling and McClure 1972). Pelleted diets should be supplemented with other foods but gibbons maintained on fruit alone develop diarrhoea (Keeling and McClure 1972). They may also be offered well-cooked meat, eggs, whole-meal bread, cheese, nuts, milk, soya beans, raw foliage, and hibiscus flowers (Badham 1967; Brody and Brody 1974). Vitamin supplements have been quite widely used.

Adult energy requirements

Assuming the energy requirements are approximately 120 kcal/d per $kg^{0.75}$, adults are likely to require about 460 kcal/d. This could be provided by 150 g of commercial primate pellets daily but a varied diet should be provided.

Growth

The growth of one captive animal from an estimated age of 1.5 to 7 years is shown in Fig. 17.1 (this curve is taken from Schultz 1944, who, prob-

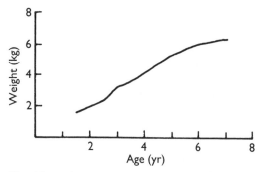

Fig. 17.1. The growth rate of an individual lar gibbon from an estimated age of 1.5 years. From Schulz (1944).

ably mistakenly assumed the animal was 3.5 years old when measurements began). Adult size is reached at approximately 6 years of age (Gittins and Raemaekers 1980). Babies hand-reared by Sasaki (1962) and Breznock *et al.* (1979) grew at about 5 g/day for the first five months of life (i.e., until weaning), but those described by Martin *et al.* (1979) had a mean daily weight gain of 2.38 g during the first six months of life. The babies hand-reared by Breznock *et al.* (1979) and anon. (1985) weighed more at weaning at 5 months of age (over 1 kg) than mother-reared babies of the same age (about 900 g).

During the second six months, growth rate averaged 3.4 g/day for one hand-reared infant and 3.8 g/day for four mother-reared infants (Breznock *et al.* 1979), and in another report of hand-rearing the average was found to be 3.84 g/day (Martin *et al.* 1979). At 1 year of age, one captive, mother-reared male weighed 2.3 kg (Brody and Brody 1974), another weighed 2.7 kg (Rumbaugh 1966).

Mother-and hand-reared animals described by Martin *et al.* (1979) did not reach 800 g until over 6 months of age.

Milk and milk intake

As far as we are aware, there are no data on the composition of gibbon milk. In view of the close taxonomic relationship to man and the comparable very slow growth rate, the composition is probably like that of human milk.

Human milk replacers, such as Enfamil (anon. 1985), Enfamil and Probana (Breznock *et al.* 1979),

Similac (Rumbaugh 1966; Badham 1967; Brody and Brody 1974) and others unspecified, (Sasaki 1962), have been used 'off-the-shelf' for hand-rearing gibbons. Martin *et al.* (1979) used a formula made from 520 g SMA powder and 210 g lactose mixed with 3 litres of water, and James (1962) used a human formula to which glucose was added. With the additional lactose, the proportions of fat, protein, carbohydrate, and ash in the formula used by Martin *et al.* (1979) were 0.23, 0.12, 0.63, and 0.01 respectively. The formula therefore contained a considerably lower proportion of fat, protein, and minerals than pure SMA. The rationale for, and advantage of, additional lactose or glucose is therefore not clear. Various workers have added supplementary paediatric vitamin drops but, here again there may be no advantage in doing this as the vitamin levels in modern human milk formulae are probably quite adequate for gibbons.

The daily milk consumption of one baby was recorded by Breznock *et al.* (1979). The intake gradually increased from about 90 ml/d during the first week to about 190 ml/d at 12 weeks of age. Broadly similar patterns of intake were reported by Sasaki (1962) and anon. (1985), but in the latter case, intake was only 43 ml on the second day. At 80 days of age the three babies reared by Martin *et al.* (1979) were all regularly taking 100 to 140 ml/d.

For the first few feeds, until the bottle-feeding technique is established, there is a case for feeding 10 per cent glucose or 10 per cent dextrose solutions rather than milk formulae because, if aspirated, these are less likely to result in pneumonia. Initial feeds should be of small volumes, for example, 7–10 ml (Sasaki 1962; Breznock *et al.* 1979), or even as little as 2–3 ml (Martin *et al.* 1979). The intake per meal of the babies reared by Breznock *et al.* (1979) gradually increased to upto 35 ml per feed at 5 months of age.

Assuming the energy density of the milk formula provided by Breznock *et al.* (1979) was 0.7 kcal/g, then the energy intake of the baby whose intake was recorded increased gradually from about 60 kcal/d during the first week to about 125 kcal/d by 12 weeks of age. This represents a relatively low daily rate of energy intake per unit

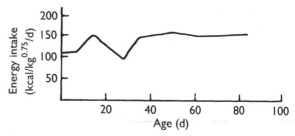

Fig. 17.2. Approximate daily energy intake provided by milk in a hand-reared gibbon per metabolic weight. From Breznock *et al.* (1979).

of metabolic weight of about 100–150 kcal/kg$^{0.75}$ (Fig. 17.2).

Lactation and weaning

By 6 weeks of age a wild infant was observed to nibble on leaves and twigs, and by 2 months of age it was able to pull vegetation to its mouth. However, it only began to ingest significant amounts of solid food by 4 months of age (Ellefson 1974). Others have suggested that weaning begins at about the third or fourth month of age, but is not complete until the end of the first year (Chivers and Raemaekers 1980; McKenna 1982), or even until 2 years of age, when the next infant is born (Ellefson 1974). In the later stages, the infant nurses only at night when the family is settled in its sleeping location (Ellefson 1974). Soft fruit pulps probably form the first weaning foods (Chivers and Raemaekers 1980).

In captivity, solid foods (baby foods and soft fruits have been used) may be introduced at about 2 to 3 months of age (Sasaki 1962; Badham 1967; Breznock *et al.* 1979; Martin *et al.* 1979; anon. 1985). Breznock *et al.* (1979) observed that hand-reared infants started showing an interest in solids at 3 months of age whereas mother-reared infants did so one month later. They recommended offering fruits, cereals, and softened monkey chow at 3 months of age.

In captivity, mother-reared infants were seen to take fruit at approximately 4 months of age (Martin *et al.* 1979).

Feeding

Human baby bottles with premature infant nipples, pet nipples or Evenflo Nipples have been used

(Brody and Brody 1974; Breznock et al. 1979; anon. 1985). During feeds, babies should be fed in an upright sitting position and 'burped' afterwards (Breznock et al. 1979). Feeding schedules have varied. It appears, however, that very young infants require feeding during the night as well as the day, approximately every 2–3 hours (James 1962; Breznock et al. 1979; Martin et al. 1979). However, one infant was fed 6 times daily from birth (Sasaki 1962). A 27-day-old infant was fed 5–6 times a day, with no feeds between 23.00 h and 07.00 h (Breznock et al. 1979). One infant that was fed on demand, required feeds around the clock until the age of 11 weeks when it slept for 7–8 hours through the night (Breznock et al. 1979).

Breznock et al. (1979) trained 4.5-month-old infants to feed themselves from hanging bottles attached to the cage over a period of 7 days. The same authors removed some captive mother-reared babies at 5 to 8 months of age and reported that they readily adapted to self-feeding from hanging bottles within 24 hours. These animals were also offered fruit, baby foods, and a soaked commercial monkey diet (Breznock et al. 1979). Martin et al. (1979) provided bottles from which hand-reared babies could drink, but observed that they did not begin self-feeding in this way until 3.5 months of age.

Accommodation

Newborn or very young gibbons have been kept in standard human incubators (Breznock et al. 1979; Martin et al. 1979; anon. 1985). The temperature was maintained 31–35 °C and 65 per cent humidity for the first 6 weeks of age. A rolled-up towel or more elaborate surrogate, such as has been devised for squirrel monkeys, should be provided for the infant to cling to in a vertical position. Some mechanism should be arranged to rock or gently shake the surrogate so that the baby is stimulated to actively cling to it. Babies should be reared within sight and sound of adult gibbons or peers.

At 2–3 months of age babies can crawl and should be moved to larger cages with facilities for climbing; initially just for a few hours each day (Breznock et al. 1979; Martin et al. 1979). From 4 months, or perhaps earlier, hand-reared infants can be kept in groups (Breznock et al. 1979).

There is very little literature on accommodation for young gibbons. Efforts should be made to provide a stimulating and varied environment, with as much contact with other gibbons as can safely be arranged.

Infant management notes

Provision of a soft, vertical, and rocking surrogate to which the baby can cling, and providing early socialization with mother-reared peers or adult gibbons are extremely important features of the management of hand-reared gibbons. Breznock et al. (1979) recommended gradual socialization with peers, but considered that early and gradual socialization beginning at 1–2 months with a non-human primate of any species would be preferable to no contact during the first 6 months.

Diapers were put on babies reared at one establishment (Martin et al. 1979). Infants should be encouraged to stand, grasp, and exercise their legs at each feeding (Breznock et al. 1979).

Physical development

At birth, babies are nearly hairless (Crandall 1945; Ellefson 1974) except for some black hair on the head (Sasaki 1962; Brody and Brody 1974). By about 1 month of age the body is covered with fur (Keeling and McClure 1972; Ellefson 1974). The eyes are open at birth (Ellefson 1974).

The deciduous dentition erupts in the following sequence. The first teeth are the middle incisors which are soon followed by the lower and then the upper lateral incisors (Schultz 1973). In one infant, however, the lower middle incisors were visible at birth (Ibscher 1967). By day 29, the first six incisors have erupted (Schultz 1973), followed by the last two at 8 weeks of age (Brody and Brody 1974). By the fourth month all incisors, the first molars, and the upper canines have erupted (Ibscher 1967). By the sixth month (Ibscher 1967), or after the seventh month (Badham 1967), the lower second molars appear after the lower canines, and by 8 months of age the full deciduous dentition is complete but there are no permanent teeth (Ibscher 1967). Rumbaugh (1966) described a similar sequence of eruption but which was completed by 5 months of age.

Behavioural development

Baby gibbons can vocalize within a few hours of birth (Ellefson 1974) and cling unaided to their mother's chests. They are never carried in the dorsal position (Crandall 1945; Badham 1967; Ellefson, 1974; Carpenter 1976).

They appear to suckle frequently during the day, and sleep at night with a nipple in the mouth (Ellefson 1974).

During the first two weeks movements on the mother are rather aimless except when rooting to nurse, but become more co-ordinated by 1 month of age (Ellefson 1974). Infants can pull themselves upright, and kneel and sit by about 6 to 8 weeks of age (Raumbaugh 1966; Brody and Brody 1974; Breznock et al. 1979). At 8 to 10 weeks babies can stand and begin to climb and by 4 months can climb skilfully and swing safely with both hands (Breznock et al. 1979). Hand-reared infants may develop more slowly but appear to be adequately co-ordinated, and able to climb and stand by 4 months of age (Rumbaugh 1966; Brody and Brody 1974; Breznock et al. 1979).

Gibbons are very dependent on their mothers for the first year, and are still carried by them on long trips until the birth of the next offspring. After that, at 2 years old, infants become gradually less well tolerated by the adults. When physical and sexual maturity are reached at about 6 years old the animals remain on the family's territorial boundaries until pairing takes place (Ellefson 1974; Gittins and Raemaekers 1980). Young gibbons in the wild are not fully independent until 7 or 8 years of age.

Stereotypical behavioural abnormalities, including thumb or wrist sucking, self-clasping, rocking, head-banging, and shaking, were seen in all four babies hand-reared by Breznock et al. (1979). However, one of these babies that was carried almost continuously by its keeper during the first 6 weeks or placed in a vertical position on a soft surrogate and frequently rocked, showed only transient thumb-sucking between 2 and 14 weeks of age. By 5 months the behaviour was apparently normal. Breznock et al. (1979) also observed that two of their animals which showed self-clasping behaviour withdrew when introduced to peers instead of showing the normal curiosity and exploratory behaviour of mother-reared infants. Infants reared under suboptimal conditions can also be very susceptible to stress, such as sudden movements, noise, handling, and unfamiliar events (Rumbaugh 1966; Breznock et al. 1979). Sasaki (1962) decribed a baby that was reared without a surrogate mother and which clenched his hands and scratched his face, presumably as a result of a frustrated drive to cling.

Disease and mortality

Survival rates for infant and juvenile *H. lar* in the field appear to be relatively high (Robbins Leighton 1987). In captivity, however, mortality rates may be high. In one report 25 (68 per cent) of 37 gibbons born in captivity to failed to survive one year (Arnold 1973). Chayet (1983), in a review of data on captive births, found that 40 per cent of 32 babies died during their first year. Pneumonia appeared to be a major cause of death in young infants surveyed by Keeling and McClure (1972).

Preventative medicine

The gibbon is thought to be susceptible to many human childhood diseases. The oral trivalent poliomyelitis vaccine has been recommended for gibbons (Keeling and McClure 1972). In addition, a family of gibbons kept in captivity in Hong Kong, were also vaccinated against small pox, cholera, typhoid, diphtheria, pertussis, and tetanus but not measles or tuberculosis (BCG) (Brody and Brody 1974). Where the risk of tuberculosis demands it, routine chest radiographs and tuberculin skin tests have also been recommended (Keeling and McClure 1972).

Hand-reared gibbons are at great risk of developing behavioural disturbances. As indicated above, surrogates and early socialization with other gibbons are of enormous importance.

Indications for hand-rearing

Brody and Brody (1974) described a procedure for providing supplementary foods (strained cereal, milk, and orange juice) to an infant whilst it clung to its own mother who was used to human intervention. Hand-rearing should be avoided if at all possible because of the high risk of behavioural disturbances. Unless facilities are excellent, and early socialization can be arranged, euthanasia should be considered. Hand-reared gibbons have proved to be 'viable sires and dams' (Breznock et al. 1979).

The indications for intervention either to hand-rear or euthanase are: if the baby is neglected or injured by the mother, or premature, sick, or failing to gain weight. The chances of another female being available for cross-fostering are extremely unlikely.

Reintegration

Martin et al. (1979) placed babies from the age of 2 months in small cages in an enclosure housing adults in order to begin socialization, and in due course the infants were released into the adult's cage. They also encouraged socialization by housing hand-reared infants with mother-reared peers whenever possible. The latter technique was also used by Breznock et al. (1979).

The social development of three juveniles released into a semi-natural environment appeared normal despite long-term caging and the lack of adult conspecifics. Socialization with peers and limited human contact during growth were considered to be relevant to the development of the normal behaviour of these animals (Paluck et al. 1970). Juveniles have been introduced into an established group of gibbons without causing great excitement or aggression. Social interactions were observed in most cases within 15 minutes (Bernstein and Schusterman 1964).

The principles for reintegration outlined for the rhesus macaque (Chapter 12) are probably equally relevant to gibbons.

References

Aldrich-Blake, F.F.G. (1978). Dispersion and food availability in Malaysian forest primates. In *Recent advances in primatology*, Vol. 1. *Behaviour* (ed. D.J. Chivers and J. Herbert), pp. 323–25. Academic Press, London.
Anon. (1985). White-handed gibbon. In Taylor, S.H. & Bietz, A.D. (Eds) *Infant diet/care notebook*. American Association of Zoo Parks and Aquariums Wheeling, Virginia.
Arnold, R.C. (1973). Births of gibbons in captivity. In *Gibbon and Siamang*, Vol. 1. (ed. D.M. Rumbaugh), pp. 221–7. Karger, Basel.
Badham, M.A. (1967). A note on breeding the pileated gibbon. *International Zoo Yearbook*, **7**, 92–3.
Berkson, G. and Chaicumpa, V. (1969). Breeding gibbons (*Hylobates lar Eutelloides*) in the laboratory. *Laboratory Animal Care*, **19**, 808–11.
Bernstein, I.S. and Schusterman, R.J. (1964). The activity of gibbons in a social group. *Folia Primatologica*, **2**, 161–70.
Breznock, A.W., Harrold, J.B., and Kawakami, T.G. (1977). Successful breeding of the laboratory-housed gibbon (*Hylobates lar*). *Laboratory Animal Science*, **27**, 222–8.
Breznock, A.W., Porter, S., Harrold, J.B. & Kawakami, T.G. (1979). Hand-rearing infant gibbons. In *Nursery care of non-human primates* (ed. G.C. Ruppenthal), pp. 287–98. Plenum Press, New York.
Brody, E.J. and Brody, A.E. (1974). Breeding Muller's Bornean gibbon. *International Zoo Yearbook*, **14**, 110–13.
Carpenter, C.R. (1963). A field study in Siam of the behaviour and social relations of the gibbon (*Hylobates lar*). In *Primate social behaviour* (ed. C.H. Southwick), pp. 17–23. Van Nostrand, New York.
Carpenter, C.R. (1976). Suspensory behaviour of gibbons, *Hylobates lar*. In *Gibbon and siamang*, Vol. 4. (ed. D.M. Rumbaugh), pp. 1–20. Karger, Basel.
Chance, M.R.A. and Jolly, C.J. (1970). *Social groups of monkeys, apes and men*. Jonathan Cape, London.
Chayet, J-M. (1983). Les gibbons. Etude zoologique et maintien en captivité. Thèse pour le Doctorat Vétérinaire. Ecole Nationale Vétérinaire D'Alfort.
Chivers, D.J. and Raemaekers, J.J. (1980). Long-term changes in behaviour. In *Malayan forest primates: ten year's study in tropical rain forest* (ed. D.J. Chivers), pp. 209–60. Plenum Press, New York.
Crandall, L.S. (1945). Our first baby gibbon is born and seems to be doing well. *Animal Kingdom*, **48**, 158–9.
Ellefson, J.O. (1974). A natural history of white-handed gibbons in the Malayan peninsula. In *Gibbon and siamang*, Vol. 3 (ed. D.M. Rumbaugh), pp. 1–136. Karger, Basel.
Flesness, N.R. (1986). Captive status and genetic considerations. In *Primates. The road to self-sustaining populations* (ed. K. Benirschke), pp. 845–6. Springer-Verlag, New York.
Fooden, J. (1971). Report on primates collected in western Thailand, Jan–April, 1967. *Fieldiana, Zoology*, **59**(1), 1–62.
Gittins, S.P. and Raemaekers, J.J. (1980). Siamang, lar and agile gibbons. In *Malayan forest primates: ten years study in tropical rain forest* (ed. D.J. Chivers), pp. 63–105. Plenum Press, New York.
Ibscher, L. (1967). Geburt und fruhe Entwicklung zweier Gibbons (*Hylobates lar* L.) *Folia Primatologica*, **5**, 43–69.
James, R.M. (1962). Hoolock gibbon (*Hylobates hoolock*). *International Zoo Yearbook*, **4**, 306.
Jones, M.L. (1968). Longevity of primates in captivity. *International Zoo Yearbook*, **8**, 183–92.
Keeling, M.E. and McClure, H.M. (1972). Clinical manage-

ment, diseases and pathology of the gibbon and siamang. In *Gibbon and siamang*, Vol. 1 (ed. D.M. Rumbach), pp. 208–15. Karger, New York.

Mackinnon, J.R. (1977). A comparative ecology of Asian apes, *Primates*, **18**, 747–72.

Mackinnon, K. (1986). The conservation status of non-human primates in Indonesia. In *Primates. The road to self-sustaining populations* (ed. K. Benirschke), pp. 99–126. Springer-Verlag, New York.

Mackinnon, J.R. and MacKinnon, K.S. (1977). The formation of a new gibbon group. *Primates*, **18**, 701–8.

Mackinnon, J.R. and Mackinnon, K.S. (1978). Comparative feeding ecology of six sympatric primates in west Malaysia. In *Recent advances in primatology*, Vol. 1. *Behaviour* (ed. D.J. Chivers and J. Herbert), pp. 305–25. Academic Press, London.

Martin, D.P., Golway, P.L., George, M.J., and Smith, J.A. (1979). Care of the infant and juvenile gibbon. In *Nursery care of non-human primates* (ed. G.C. Ruppenthal), pp. 299–305. Plenum Press, New York.

McKenna, J.J. (1982). The evolution of primate societies, reproduction and parenting. In *Primate behaviour* (ed. J.L. Fobes and J.E. King), pp. 87–133. Academic Press, London.

Napier, J.R. and Napier, P.H. (1967). *A handbook of living primates*. Academic Press, New York.

Paluck, R.J., Lieff, J.D., and Esser, A.H. (1970). Formation and development of a group of juvenile *Hylobates lar*. *Primates*, **11**, 189–94.

Raemaekers, J.J. (1978*a*). Changes through the day in the food choice of wild gibbons. *Folia Primatologica*, **30**, 194–205.

Raemaekers, J.J. (1978*b*). Competition for food between lesser apes. In *Recent advances in primatology*, Vol. 1. *Behaviour* (ed. D.J. Chivers and J. Herbert), pp. 327–36. Academic Press, London.

Raemaekers, J.J. (1979). Ecology of sympatric gibbons. *Folia Primatologica*, **31**, 227–45.

Robbins Leighton, D. (1987). Gibbons: territoriality and monogamy. In *Primate societies* (ed. B.B. Smuts, D.L. Cheney, R.M. Seyfarth, R.W. Wrangham, and T.T. Struhsaker), pp. 135–45. University of Chicago Press.

Rumbaugh, D.M. (1966). The behaviour and growth of a lowland gorilla and gibbon. *Zoonooz* (Newsletter of the Zoological Society of San Diego), **40**, 12–18.

Sasaki, T. (1962). Hand-rearing a baby gibbon (*Hylobates lar*). *International Zoo Yearbook*, **4**, 289–90.

Schultz, A.H. (1944). Age changes and variability in gibbons. A morphological study on a population sample of a man-like ape. *American Journal of Physical Anthropology*, **21**, 1–129.

Schultz, A.H. (1973). The skeleton of the hylobatidae and other observations on their morphology. In *Gibbon and siamang*, Vol. 2 (ed. D.M. Rumbaugh), pp. 1–54. Karger, Basel.

Wolfheim, J.H. (1983). *Primates of the world*, pp. 661–5. University of Washington Press, Seattle.

Zuckerman, S. (1981). *The social life of monkeys and apes*. Routledge & Kegan Paul, London.

Chimpanzee

18 Chimpanzee

Species

The chimpanzee *Pan troglodytes*

ISIS No.   1406009002001

Status, subspecies, and distribution

Three subspecies are recognized: *P.t. verus*, the western chimpanzee; *P.t. troglodytes*, the central chimpanzee; and *P.t. schweinfurthi*, the eastern chimpanzee. Lee *et al.* (1988) and the IUCN (1990) classified the western subspecies, which occurs from Senegal to Nigeria, as endangered, and the other two which are found from Nigeria to the Zaïre river and east of the Zaïre river respectively, as vulnerable. These authors estimated that the total wild population might be 200 000 but accurate census data are not available for many parts of the range.

The captive population is about 1770, of which 742 are known captive-born and 425 are known wild-caught (Seal and Flesness 1986). Chimpanzees are widely kept in zoos and for research. In 1984, Johnsen and Whitehair (1986) estimated the number of actual and potential breeders in principal US research institutions to be 1039.

Sex ratio

The sex ratio at birth is close to 1:1 (Sugiyama and Koman 1979). Of 214 births in captivity, 110 were male and 104 female (Bourne 1972).

Social structure

In the wild, social groups of chimpanzees are fluid and group compositions vary daily. Territories are not defended (Reynolds 1979). Groups of two to nine are most commonly observed (Goodall 1965; V. Reynolds and F. Reynolds 1965a), and these can be categorized into four types: mother groups, adult male groups, mixed-sex adult groups, and mixed-sex groups of all ages (Reynolds 1963, 1965; Goodall 1965; V. Reynolds and F. Reynolds 1965b).

Breeding age

Swelling of the sexual skin usually begins at 6 to 7 years and the onset of menarche follows this about 18 months later. There is then typically an approximately 18 month period in which the females show oestrous cycles but do not conceive (Graham 1970; Keeling and Roberts 1972; Bourne 1972). First breeding therefore usually occurs at between 8 and 12 years, but has been recorded in 5-year-olds (Seal and Flesness 1986). Menopause has not been recorded and breeding has been observed in a female of 47 years (Keeling and Roberts 1972; Bourne 1972).

Spermatogenesis is thought to begin at 7 to 9 years of age (Keeling and Roberts 1972), but males of just 6 years old have sired young (Seal and Flesness 1986).

Longevity

The life expectancy in the wild is thought to be about 30 years (Reynolds 1979). In captivity, some have survived over 40 years (Keeling and Roberts 1972), and there is one report of an animal living to about 60 or 70 years (Goodall 1965).

Seasonality

Chimpanzees breed throughout the year. The duration of the oestrous cycle is 33 to 38 days (Graham 1970; Bourne 1972). The pre-swelling phase lasts about a week and the swelling, tumescent phase lasts 18 days. The swelling reaches maximum size after 15 to 20 days when receptivity is also highest. Ovulation occurs during the next 5 to 6 days whilst the swelling remains maximal. The swelling then decreases over about 10 days but may occur very suddenly over 48 hours. The cycle is completed with the menstrual phase when visible blood flow lasts about 3 days (Keeling and Roberts 1972).

Cycling is resumed about 1 year post-partum (Sugiyama and Koman 1979), although some cyclicity of swelling may occur without menstruation during lactation (Keeling and Roberts 1972). Interbirth intervals are generally 3 to 4 years, but vary considerably and are much less if the infant does not survive (Reynolds and Reynolds, 1965b; Sugiyama and Koman 1979; Adang et al. 1987).

Gestation

When the gestation period was calculated from the last day of maximum swelling, a mean gestation of 228 days was calculated (Graham et al. 1985), with a range of 194–250 days. A slightly longer estimate (238–244 days) was obtained from dates of artificial insemination (Hardin et al. 1975).

Pregnancy diagnosis

The oestrous cycle ceases after conception, although sexual swelling and some menstrual blood may be observed for one or two cycles (Keeling and Roberts 1972). Mammary development and abdominal swelling in the later stages of pregnancy may not be marked. Pregnancy can, with experience, be diagnosed by palpation of the uterus as early as 30 days after conception (Mahony 1980; McClure et al. 1973a). Ultrasonography can be used to detect pregnancy, and the fetal skeleton becomes visible by radiography at about 100 days after conception (Keeling and Roberts 1972; McClure et al. 1973a). Chorionic gonadotrophins can be detected in the urine using either standard test kits for humans or for non-human primates (Keeling and Roberts 1972; McClure et al. 1973a; Hardin et al. 1975; Mahony 1980).

Birth

Births occur at any hour of the day or night (Bo 1971). Labour lasts from 1 to 8 hours (Yerkes 1943; Bo 1971; Keeling and Roberts 1972) and its onset is marked by restlessness, frequent changes of position, irregular breathing, and contractions. The actual delivery is usually rapid (Budd and Smith 1943; Keeling and Roberts 1972). The normal presentation is cranial and the incidence of dystocia is low (Yerkes 1943). The umbilical cord is severed by the mother before or after delivery of the placenta, and the placenta is usually passed 45 to 60 minutes after the baby is born (Budd and Smith 1943; Keeling and Roberts 1972). Experienced mothers may suck the babies nostrils and mouth and clean it all over. Babies usually pass meconium within a few hours of birth (Budd and Smith 1943).

Litter size

The usual litter size is 1, but twins occur with a frequency of about one per 90 births. Triplets occur rarely (Yerkes 1943; Yamamoto 1967; Bourne 1972; Keeling and Roberts 1972; Seal et al. 1985).

Adult weight

Adult males weigh between 32 and 60 kg with an average of 45 to 50 kg, and adult females have a range from about 31 to 55 kg with an average of about 40 kg (Schultz 1940; Yerkes 1943).

Neonate weight

Birth weights are typically between 1.0 and 2.6 kg (Schultz 1940; Budd and Smith, 1943; Yerkes 1943; Joy 1971; Keeling and Roberts 1972; Hardin et al. 1975; Graham et al. 1985). Average birth weights have been recorded at 1.58 kg (Yerkes 1943) and 1.8 to 1.9 kg (Schultz 1940; Graham et al. 1985). There is no apparent difference between the sexes. Twins that were successfully hand-reared weighed 800 g and 850 g (Yamamoto 1967).

Adult diet

In the wild, the diet consists mainly of fruit, but leaves, blossoms, bark and seeds, and also insects are eaten. Chimpanzees also hunt and eat mammals such as other primates, bush pigs, and a variety of other small animals (Goodall 1965; Telecki 1973; Reynolds 1979). The diets fed to captive chimpanzees were reviewed by Bourne (1971). Often the animals are fed on commercially available pelleted primate diets supplemented with a variety of fresh fruit and fresh and cooked vegetables, and a small amount of cooked lean meat and milk.

Adult energy requirements

The energy requirement of captive adult chimpanzees has been estimated to be 2000–3000 kcal/

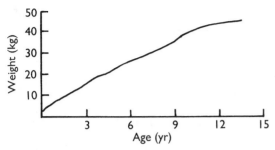

Fig. 18.1. Typical growth curve of the chimpanzee. From Schulz (1940) and Gavan (1970).

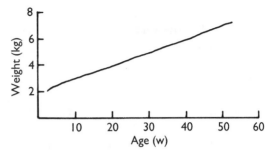

Fig. 18.2. Typical growth curve of infant hand-reared chimpanzees. From Budd and Smith (1943) and Lawrence and Gorzitze (1985).

d (Yerkes 1943), which is about 120–160 kcal/d per $kg^{0.75}$. The energy metabolism of the species is thought to be similar to that of man (Dale *et al.* 1970).

Growth

A typical growth curve based on data from Gavan (1971) and Schultz (1940) is shown in Fig. 18.1. The weight gain of the two sexes is similar to 8 years when a weight of about 32 kg is reached. At about this age there is an increase in growth rate specially in males (Smith *et al.* 1975). The adult weight of 40 to 50 kg is reached at about 12 years of age (Schultz 1940; Gavan 1971; Lawrence and Gorzitze 1985). During the first year, growth is approximately linear, but there is considerable variation in growth rates among individuals, and hand-reared infants tend to grow more rapidly than those reared by their mothers (Davenport and Rogers 1970; McClure *et al.* 1973b; J. Fritz and P. Fritz 1982; Lawrence and Gorzitze 1985). The typical growth pattern of hand-reared chimpanzees, based on data from Lawrence and Gorzitze (1985) and Budd and Smith (1943) is shown in Fig. 18.2. Babies may lose weight in the immediate neonatal period (for example, Carmichael *et al.* 1961).

Milk and milk intake

The composition of the milk has been reported by Ben Shaul (1962b), Bourne (1971, 1972), and Buss (1971). The composition of the fresh milk was found by Buss (1971) to be approximately 3.0 per cent fat, 1.0 per cent protein, 7.1 per cent lactose, and 0.21 per cent ash. The dry matter composition would therefore have been 27 per cent, 9 per cent and 63 per cent, fat, protein, and lactose respectively, and the energy density would have been about 0.6 kcal/ml. This is similar to the composition of human milk, and human milk replacers are recommended (Bourne 1972; Keeling and Roberts 1972; Fritz *et al.* 1985; Lawrence and Gorzitze 1985). Although these milk substitutes have often been supplemented with additional vitamins there is probably no need for this. Chimpanzees have also been successfully reared on a formula based on cow's milk (Yerkes 1943).

Buss (1971) suggested that the milk energy intake of chimpanzee babies was similar to that of human infants at 120–140 kcal/kg per day. However, Fritz *et al.* (1985) recommended that an average of 100 ml of human milk replacer per kilogram body weight should be consumed each day. The latter would be consistent with about 70 kcal/kg per day, or about 90 kcal/d per $kg^{0.75}$. This probably represents a minimum and may not sustain growth because the maintenance requirement is unlikely to be substantially less than this.

Lactation and weaning

Observations in the wild suggested that suckling bouts are initiated by the mother lifting the baby to her breasts (Goodall 1965). It can take inexperienced mothers several days before they allow regular suckling (Rogers and Davenport 1970; Keeling and Roberts 1972). In wild chimpanzees, nipple contact was recorded an average of 2.7 times per hour, for 2.5 minute bouts during the first 6 months (Fritz and Fritz 1982).

In captivity lactation can last into the third year (Yerkes 1943). In the wild and in captivity, solid foods are first taken at 4 to 6 months of age (Fritz and Fritz 1982; Fritz et al. 1985).

Feeding

Until the sucking response is well established, which may take a few days (Mason 1970), or for the first few feeds, hand-reared babies should be fed on 10 per cent glucose solution which reduces the risk of pneumonia should some aspiration occur (Fritz et al. 1985). Bottles and nipples designed for human babies are suitable for feeding infant chimpanzees.

The feeding schedules reported in the literature vary. Some have fed every 2 to 3 hours in the early stages (Budd and Smith 1943; Yamamoto 1967), but feeding at 4 hourly intervals throughout the night and day is adequate (Yerkes 1943; Bourne 1971; Lawrence and Gorzitze 1985). The frequency of bottle-feeding is gradually reduced to 3 or 4 per day by 1 year of age. It has been suggested that decreasing the number of feeds too early may lead to behavioural abnormalities (Fritz et al. 1985). In the wild, sucking is frequent as mentioned above (see Lactation and weaning).

A survey of various hand-rearing protocols has shown that the age at which other foods are offered has varied considerably. Rice cereal has been added to milk as early as 14 days, and babies have been weaned on to a mixture of blended monkey chow and evaporated milk at 5 to 6 months of age (Lawrence and Gorzitze 1985). Others have first offered solids at a later stage. There is little advantage in giving solid foods before 4 months of age, and addition of solids to the diet too early may be harmful (Fritz et al. 1985). The first foods introduced to the infant's diet have been rice cereal, fruit mixtures, and monkey chow blended in to the milk. The latter may not be most suitable; at this stage foods should be easily digestible.

Accommodation

Incubators designed for human babies are ideal for newborn chimpanzees (Yamamoto 1967; Keeling and Roberts 1972; Bowen 1985; Fritz et al. 1985). Incubators should be kept out of direct sunlight and away from draughts (Alford 1985). Once the babies are able to maintain stable body temperature (a rectal temperature of 37 °C — Bowen 1985) they can be kept in a less constant environment with temperatures in the 20 to 30 °C range (Fritz and Fritz 1985).

Infant management notes

It is accepted that the physical and social environment during growth, particularly in the early stages, have a major effect on the behavioural development of chimpanzees. There has, however, been little research on the behavioural consequences of differing early environments in this species. Guidelines that have been proposed for attempting to hand-rear chimpanzees that will subsequently be able to reproduce and care for their infants (for example, Fritz and Fritz 1985), have therefore been on knowledge of the subject in other species (see the section on rhesus macaques, chapter 12) and experience in caring for young chimpanzees. Fritz and Fritz (1985) stress the time and commitment required for hand-rearing chimpanzees, and the great responsibility of the human foster-parent in taking on this role and assessing and overseeing the development of the baby. The reader is referred to their guidelines on this subject, which is very important to the animals' welfare.

Infant chimpanzees stay with their mothers for at least 3 to 4 years and are largely dependent on them for at least the first year. Separation and independence occur gradually. In hand-rearing, there is a long-term requirement for providing a stable, caring, and stimulating environment, and a gradual transition to independence.

The umbilicus of babies taken for hand-rearing soon after birth should be ligated and cut 5 to 7.5 cm from the abdomen and cleaned with an iodine preparation (Keeling and Roberts 1972; Alford 1985).

During the first 8 months urination was observed to occur 8 times daily, and the frequency of defaecation decreased from 7 times a day in the first week to 4 times a day at 6 months (Yamamoto 1967). Diapers are commonly used on hand-reared infants and need changing frequently. The perineum should be kept clean using wipes or warm-water washes. If diapers are not used, cage cleaning has to be frequent.

A self-feeding system available is not advisable because it reduces the time of contact between the infant and keeper and may result in behavioural abnormalities (Fritz and Fritz 1985). Hand-reared babies have shown a slower rate of development of motor skills than those reared by their mothers. It is thought that the amount of stimulation (tactile, kinaesthetic, visual, and auditory) the infant experiences during the first year is crucial to its development. Provision of an interesting and stimulating environment is therefore recommended (Fritz and Fritz 1985).

Early contact with peers or infants of similar age is considered beneficial for normal behavioural development. However, peers do not act as substitute mothers and the presence of a trusted human care-giver remains essential for this role (Fritz and Fritz 1985).

Physical development

The eyes are open at birth. The head is covered with hair but the ventrum is nearly naked (Budd and Smith 1943). A tuft of long white hairs is present over the rump for the first 2 to 3 years. Adolescent females have long cheek hair which is less noticeable in males (Goodall 1965).

The sequence of eruption of the deciduous teeth is as follows: central incisors, 2–4 months; lateral incisors, 3–4.5 months; upper first molars, 3.7–5 months; lower first molars, 4.6–8 months; second molars, about 4 months after the first molars; and canines 13–16 months. The permanent teeth develop as follows: upper and lower first molars, 33–36 months; lower first incisors, 57–70 months; upper first incisors, 61–72 months; lower second incisors, 60–73 months; upper second incisors, 64–74 months; lower first premolars, 61–80 months; upper first premolars, 69–79 months; upper and lower second premolars, 72–74 months; upper and lower second molars, 73–82 months; upper and lower canines 81–99 months; lower third molars, 104–121 months; upper third molars, 118–134 months (Schultz 1940; Budd and Smith 1943). The dentition is therefore complete at about 11 years of age.

Behavioural development

At birth, baby chimpanzees are unable to cling without support. They are carried and cradled by their mothers and are unable to move about on them until about 3 months old. At 6 months old they can support their bodies and crawl, and begin to make short forays away from their mothers (Reynolds 1979). Postural and early locomotory development has been descibed by Riesen and Kinder (1952). By the end of the first year babies have begun to climb (Yerkes 1943), and to play with other members of the group, but remain under close maternal supervision. By 2 years old babies can feed themselves and move independently, but are still often carried (Goodall 1965; Reynolds 1979). By 3 to 4 years of age infants are practically independent in feeding and locomotion but stay close to their mothers and still ride on them at times. Juveniles build their own nests and sleep apart from their mothers by 4 to 5 years of age (Reynolds 1979).

The conditions under which the baby is reared influence the rate and quality of the behavioural development. Isolation results in depression, apathy, and neurosis (Reynolds and Reynolds 1965a). A high incidence of abnormal behaviour such as hand-clapping, foot-stamping, rocking, and other stereotypical patterns, has been observed in hand-reared infants, as well as abnormal timidity and lack of adaptibility (Davenport and Rogers 1970).

Fritz and Fritz (1985) have compiled schedules of the stages and landmarks of behavioural development of mother-reared and hand-reared chimpanzees (see also Riesen and Kinder 1952), and have listed indicators of pathological behavioural development during the first year of life. This information may help in assessing the behavioural development of hand-reared chimpanzees.

Disease and mortality

Seal et al. (1985) and Seal and Flesness (1986) have reviewed infant mortality in captive-born chimpanzees. They found that the overall mortality rate during the first year was about 20 per cent, about half of which occurred during the first 24 hours, and that during the second year mortality was 2.7

per cent. The rate of mortality amongst twins is higher (Keeling and Roberts 1972; Seal et al. 1985). Bourne (1972) found a similar first year mortality rate (about 20 per cent) among 214 births.

The causes of mortality are various, but include abnormal maternal behaviour (Adang et al. 1987), trauma at birth, adverse environmental conditions, and infections (Seal et al. 1985). Bowen (1985) reported susceptibility to streptococcal meningitis amongst neonates, and that pneumonia is mainly seen in the 2–6-month-old age group.

Survival rates are higher in babies born to multiparous mothers compared with those born to primipara, and there is a five-fold greater risk of mortality for a baby born to a mother whose previous baby was stillborn or died on the day of birth (Seal et al. 1985).

Preventative medicine

Institutions vary in their vaccination regimes for young great apes. Vaccines are available against the human diseases of diphthteria (*Corynebacterium diphtheriae*), tetanus (*Clostridium tetani*), whooping cough (*Bordetella pertussis*), measles and mumps (paramyxovirus infections), rubella (togavirus), and poliomyelitis (picornavirus). Great apes are susceptible to polio, likely to be susceptible to tetanus, and although infections by the other organisms occur, they tend not to cause clinical disease (Loomis 1985).

The trivalent oral poliomyelitis vaccine appears to be effective, and is safe and has been widely used in great apes (Allmond et al. 1967; Loomis 1985; Hunt 1986). In view of this and the susceptibility to the disease there is quite a strong argument for this vaccination (Loomis 1985). The regime recommended for children has been employed: administration of doses at 2, 4, and 18 months followed by a booster at 5 years.

Loomis (1985) also considered that it may be advisable to vaccinate against tetanus using toxoid prepared for human administration.

The case for the other vaccinations is not so clear. There is insufficient information for assessing the cost of the benefits, or the risks of vaccinating versus not vaccinating, although vaccination against measles appears to be safe and free of side-effects in chimpanzees (Hunt 1986).

Balantidium coli may cause enteritis when a young hand-reared chimpanzee becomes infected with this protozoan when first reintroduced to carriers. This is difficult to avoid and the animal usually controls the infection.

Indications for hand-rearing

In view of the difficulties of providing a suitable environment for normal behavioural development, babies should not be taken for hand-rearing without careful consideration. Inexperienced mothers may take a few days before establishing good maternal behaviour and it is better to monitor the babies' status carefully and to avoid intervening at an early stage (Keeling and Roberts 1972). If the mother is ill or if she completely rejects the baby it has to be removed for hand-rearing, euthanasia, or for cross-fostering. The latter has been reported in chimpanzees (Van Wulffen Palthe and Van Hooff 1975; Adang et al. 1987).

Reintegration

Young chimpanzees can be more easily reintroduced to established groups than adults (Adang et al. 1987). However, the success depends upon the behavioural quality of the infant and thus on its previous environment and upbringing. Infants hand-reared in an impoverished environment may show a variety of abnormal behaviours including social behaviour (Reynolds 1965; Davenport and Rogers 1970; Fritz and Fritz 1985). For reintegration it is essential that social development is encouraged from an early stage, first by providing a high quality of foster-care, then by exposure to peer groups (Reynolds and Reynolds 1965a; Fritz and Fritz 1985). Before the infant is placed in with the group the animals should be familiar with each other through adjacent caging, and escape routes should be provided when the reintroduction is made (Kortland 1960). It may be necessary for the care-giver to accompany the young animal as far as is possible, during the initial reintroduction period (Fritz 1986).

References

Adang, O.M.J., Wensing, J.A.B., and Van Hoof, J.A.R.A.M. (1987). The Arnhem Zoo colony of chimpanzees *Pan troglodytes*: development and management techniques. *International Zoo Yearbook*, **26**, 236–48.

Alford, P.L. (1985). The prevention and control of infectious disease in the nursery. In *Clinical management of infant great apes* (ed. C.E. Graham and J.A. Bowen), pp. 35–42. Alan R. Liss, New York.

Allmond, B.W., Froeschle, J.E., and Guilloud, N.B. (1967). Paralytic poliomyelitis in large laboratory primates; virologic investigation and report on the use of oral poliomyelitis virus (OPV) vaccine. *American Journal of Epidemiology*, **85**, 229–39.

Ben Shaul, D.M. (1962). The composition of the milk of wild animals. *International Zoo Yearbook*, **4**, 333–45.

Bo, W.J. (1971). Parturition. In *Comparative reproduction of nonhuman primates* (ed. E.S.E. Hafez), pp. 302–14. Charles C. Thomas, Springfield, Illinois.

Bourne, G.H. (1971). Nutrition and diet of chimpanzees. In *Breeding primates* (ed. W.I.B. Beveridge), pp. 373–400. Karger, Basel.

Bourne, G.H. (1972). Breeding chimpanzees and other apes. In *Breeding primates* (ed. W.I.B. Beveridge), pp. 24–33. Karger, Basel.

Bowen, J.A. (1985). Intensive care of infant chimpanzees. In *Clinical management of infant great apes* (ed. C.E. Graham and J.A. Bowen), pp. 35–42. Alan R. Liss, New York.

Budd, A. and Smith, L.G. (1943). On the birth and upbringing of the female chimpanzee 'Jacqueline'. *Proceedings of the Zoological Society of London, Series A*, **113**, 1–20.

Buss, D.H. (1971). Mammary glands and lactation. In *Comparative reproduction of nonhuman primates* (ed. E.S.E. Hafez), pp. 315–33. Charles C. Thomas, Springfield, Illinois.

Carmichael, L., Kraus, M.B., and Reed, T. (1961). The Washington National Zoological Park infant, Tomoko. *International Zoo Yearbook*, **3**, 88–93.

Dale, H.E., Shanklin, M.D., Johnson, H.D., and Brown, W.H. (1970). Energy metabolism in the chimpanzee. In *The chimpanzee*, Vol. 2 (ed. G.H. Bourne), pp. 100–22. Karger, Basel.

Davenport, R.K. and Rogers, C.M. (1970). Differential rearing of the chimpanzee, a project survey. In *The chimpanzee*, Vol. 2 (ed. G.H. Bourne), pp. 183–220. Karger, Basel.

Fritz, J. (1986) Resocialisation of asocial chimpanzees. In *Primates. The road to self-sustaining populations* (ed. K. Benirschke), pp. 351–9. Springer-Verlag, New York.

Fritz, J. and Fritz, P. (1982). Great ape hand-rearing with a goal of normalcy and a reproductive continuum. *Proceedings of the American Association of Zoo Veterinarians*, 27–31.

Fritz, J. and Fritz, P. (1985). The hand-rearing unit: management decisions that may affect chimpanzee development. In *Clinical management of infant great apes* (ed. C.E. Graham and J.A. Bowen) pp. 1–34. Alan R. Liss, New York.

Fritz, J., Ebert, J.W., and Carland, J.F. (1985). Nutritional management of infant great apes. In *Clinical management of infant great apes* (ed. C.E. Graham, and J.A. Bowen), pp. 141–56. Alan R. Liss, New York.

Gavan, J.A. (1971). Longitudinal post-natal growth in the chimpanzee. In *The chimpanzee*, Vol. 4 (ed. G.H. Bourne), pp. 46–102. Karger, Basel.

Goodall, J. (1965). Chimpanzees of the Gombe Stream Reserve. In *Primate behaviour: field studies of monkeys and apes*, Vol. 1 (ed. I. Devore), pp. 425–73. Holt, Reinhart & Winston, New York.

Graham, C.E. (1970). Reproductive physiology of the chimpanzee. In *The chimpanzee*, Vol. 3 (ed. G.H. Bourne), pp. 183–270. Karger, Basel.

Graham, C.E., Bown, J.A., Billhymer, B., and Cummins, L.B. (1985). Fetal maturity estimation by lecithin/sphingomyelin ratios, pregnancy duration and caesarian section in chimpanzees. In *Clinical management of infant great apes* (ed. C.E. Graham and J.A. Bowen), pp. 35–42. Alan R. Liss, New York.

Hardin, C.J., Liebherr, G., and Fairchild, O. (1975). Artificial insemination in chimpanzees *Pan troglodytes*. *International Zoo Yearbook*, **15**, 132–4.

Hunt, R.D. (1986). Viral diseases of neonatal and infant nonhuman primates. In *Primates. The road to self-sustaining populations* (ed. K. Benirschke), pp. 725–42. Springer-Verlag, New York.

IUCN (1990). *1990 IUCN red list of threatened animals*, pp. 13–14. IUCN, Gland, Switzerland.

Johnsen, D.O. and Whitehair, L.A. (1986). Research facility breeding. In *Primates. The road to self-sustaining populations* (ed. K. Benirschke), pp. 499–511. Springer-Verlag, New York.

Joy, J.E. Jr. (1971). Pregnancy toxaemia in a multiparous chimpanzee *Pan troglodytes troglodytes* at Dallas Zoo. *International Zoo Yearbook*, **11**, 239–41.

Keeling, M.E. and Roberts, J.R. (1972). Breeding and reproduction of chimpanzees. In *The chimpanzee*, Vol. 5 (ed. G.H. Bourne), pp. 127–52. Karger, Basel.

Kortland, A. (1960). Can lessons from the wild improve the lot of captive chimpanzees? *International Zoo Yearbook*, **2**, 76–80.

Lawrence, W.A. and Gorzitze, A.B. (1985). Assessment of postnatal weight gain in nursery-reared infant chimpanzees. In *Clinical management of infant great apes* (ed. C.E. Graham and J.A. Bowen), pp. 157–64. Alan R. Liss, New York.

Lee, P.C., Thornback, J., and Bennett, E.L. (1988). *Threatened primates of Africa*, pp. 106–22. IUCN, Gland, Switzerland.

Loomis, M.R. (1985). Immunoprophylaxis in infant great apes. In *Clinical management of infant great apes* (ed. C.E. Graham and J.A. Bowen), pp. 107–12. Alan R. Liss, New York.

Mahony, C.J. (1980). Breeding chimpanzees for biomedical research: a nine-year evaluation. In *Non-human primate models for study of human reproduction* (ed. T.C. Anand Kumar), pp. 159–68. Karger, Basel.

Mason, W.A. (1970). Chimpanzee social behaviour. In *The*

chimpanzee, Vol. 2 (ed. G.H. Bourne), pp. 265–88. Karger, Basel.

McClure, H.M., Guilloud, N.B., and Keeling, M.E. (1973*a*). Clinical pathology data for the chimpanzee and other anthropoid apes. In *The chimpanzee*, Vol. 6 (ed. G.H. Bourne), pp. 121–81. Karger, Basel.

McClure, H.M., Pieper, W.A., Keeling, M.E., Jacobson, C.B., and Schlant, R.C. (1973*b*). Down's-like syndrome in a chimpanzee. In *The chimpanzee*, Vol. 6 (ed. G.H. Bourne), pp. 182–214. Karger, Basel.

Reynolds, V. (1963). An outline of the behaviour and social organisation of forest-living chimpanzees. *Folia Primatologica*, **1**, 95–102.

Reynolds, V. (1965). *Budongo, a forest and its chimpanzees*. Methuen, London.

Reynolds, V. (1979). Some behavioural comparisons between the chimpanzee and the mountain gorilla. In *Primate ecology; problem-oriented field studies* (ed. R.W. Sussman), pp. 323–39. Wiley, New York.

Reynolds, V. and Reynolds, F. (1965*a*). The natural environment and behaviour of chimpanzees *Pan troglodytes schweinfurthi* and suggestions for their care in zoos. *International Zoo Yearbook*, **51**, 141–4.

Reynolds, V. and Reynolds, F. (1965*b*). Chimpanzees of the Budongo Forest. In *Primate behaviour: field studies of monkeys and apes*, Vol. 1 (ed. I. Devore), pp. 368–424. Holt, Reinhart & Winston, New York.

Riesen, A.H. and Kinder, E.F. (1952). *Postural development of infant chimpanzees*. Yale University Press, New Haven.

Rogers, C.M. and Davenport, R.K. (1970). Chimpanzee maternal behaviour. In *The chimpanzee*, Vol. 3 (ed. G.H. Bourne), pp. 361–8. Karger, Basel.

Schultz, A.H. (1940). Growth and development of the chimpanzee. *Contributions to embryology*. Publication No. 578, Carnegie Institution, Washington.

Seal, U.S. and Flesness, N.R. (1986). Captive chimpanzee populations past, present, and future. In *Primates. The road to self-sustaining populations* (ed. K. Benirschke), pp. 47–55. Springer-Verlag, New York.

Seal, U.S., Flesness, N., and Foose, T. (1985). Neonatal and infant mortality in captive-born great apes. In *Clinical management of infant great apes* (ed. C.E. Graham and J.A. Bowen), pp. 193–203. Alan R. Liss, New York.

Smith, A.M., Butler, T.M., and Pace, N. (1975). Weight of colony-reared chimpanzees. *Folia Primatologica*, **35**, 1–29.

Sugiyama, Y. and Koman, J. (1979). Social Structure and dynamics of wild chimpanzees at Bossou, Guinea. *Primates*, **20**, 323–9.

Telecki, G. (1973). *The predatory behaviour of wild chimpanzees*. Bucknell University Press, Lewisburg.

Van Wulffen Palthe, T. and Van Hooff, J.A.R.A.M. (1975). A case of the adoption of an infant chimpanzee by a suckling foster chimpanzee. *Primates*, **16**, 231–4.

Yamamoto, S. (1967). Notes on hand-rearing chimpanzee twins *Pan troglodytes* at Kobe Zoo. *International Zoo Yearbook*, **71**, 97–8.

Yerkes, R.M. (1943). *Chimpanzees. A laboratory colony*. Yale University Press, New Haven.

List of products

Abidec	*Multivitamin solution.* Parke-Davis Veterinary, Usk Road, Pontypool, Gwent NP4 OYH, UK
Aptamil	*Human milk replacer.* Milupa Ltd, Milupa House, Uxbridge Road, Hillingdon, Middlesex UB10 ONE, UK
Casilan	*Milk protein concentrate.* Glaxo-Farley Foods, Glaxo Laboratories Ltd, Torr Lane, Plymouth PL3 5UA, UK
Complan	*Vitamin and mineral fortified drink mix.* Farley Health Products Ltd, Nottingham NG2 3AA, UK.
Cytacon	*Vitamin B_{12} syrup.* Duncan, Flockhart & Co Ltd, 700 Oldfield Lane North, Greenford, Middlesex UB6 OHD, UK
Dextri-Maltose (R)	Mead Johnson Nutritional Group 2400 West Lloyd Expressway, Evansville, Indiana 47721–3331, USA
Enfamil	*Human milk replacer.* Mead Johnson Nutritional Group, 2400 West Lloyd Expressway, Evansville, Indiana 47721–3331, USA.
Evenflo Nipples	Evenflo Products Company, 771 North Freedom Street, Ravenna, Ohio 44266, USA
Farex	*Baby rice cereal.* Farley Health Products Ltd, Nottingham NG2 3AA, UK
Ionalyte	*Electrolyte solution.* Intervet UK Ltd, Science Park, Milton Road, Cambridge CB4 4FP, UK
Lactogen	*Human milk replacer.* Nestlé Co Ltd, St George's House, Croydon CR9 INR, UK
Lactol	*Puppy milk formula.* Sherley's Division. Ashe Laboratories Ltd, Leatherhead, Surrey, KT22 7JZ, UK
Lamlac	*Sheep milk replacer.* Volac Ltd, Orwell, Royston, Hertfordshire, SG8 5QX, UK
Mazuri	*Primate diet.* Special Diets Services Division of BP Nutrition UK Ltd, P.O. Box 705, Stepfield, Witham, Essex CM8 3AB, UK
Meritene	*Nutrient powder* Sandoz Nutrition, 5100 Gamble Drive, St Louis Park, Minnesota 55416, USA
Paladec	*Multivitamin preparation.* Warner Lambert Company Parke Davis 2800 Plymouth Road, AM Arbor, Michigan 48105, USA
Premature Infant Formula	Mead Johnson Nutritional Group 2400 West Lloyd Expressway, Evansville, Indiana 47721–3331, USA.
Primilac	*Milk replacer formula for non-human primates.* Bio-Serve Inc. PO Box 450, Frenchtown, New Jersey 08825, USA
Probana	Mead Johnson Nutritional Group 2400 West Lloyd Expressway, Evansville, Indiana 47721–3331, USA.
Purina Monkey Chow	Ralston Purina Manufacturing Co, Checkerboard Square, St. Louis, Missouri 63188, USA

Rovimix	*Vitamin supplements.* Hoechst UK Ltd, Walton Manor, Walton, Milton Keynes, Buckinghamshire MK7 7AJ, UK
Similac	*Human milk replacer.* Ross Laboratories, 625 Cleveland Avenue, Columbus, Ohio 43216, USA
Similac with Iron	*Human milk replacer.* Ross Laboratories, 625 Cleveland Avenue, Columbus, Ohio 43216, USA
SMA	*Human milk replacer.* Wyeth Laboratories, Huntercombe Lane South, Taplow, Maidenhead, Berkshire SL6 0PH, UK
Sustagen	*Powdered food.* Mead Johnson Nutritional Group, 2400 West Lloyd Expressway, Evansville, Indiana 47721–3331, USA
Vi-Daylin-M	*Vitamin syrup.* Ross Laboratories, 625 Cleveland Avenue, Columbus, Ohio 43216, USA
Vionate	*Vitamin/mineral supplement.* Ciba–Geigy Agrochemicals, Whittlesford, Cambridge CB2 4QT, UK
ZF6	*Canned carnivore diet.* Spillers Dalgety, New Malden House, 1 Blagdon Road, New Malden, Surrey KT3 4TB, UK
Zu Preem Marmoset Diet	Hill's Pet Products Inc. P.O. Box 148, Topeka, Kansas 66601, USA

Index

abortion 15, 44, 49, 54, 62, 76–7, 85, 90, 95–6, 103
anaemia 66
Anthropoidea 3
Aotus trivirgatus 3, **65–9**
'aunting' behaviour 85, 96, 129

baboon
   common, savannah, yellow, or olive, *see Papio cynocephalus*
   Gelada, *see Theropithecus gelada*
   Hamadryas, *see Papio hamadryas*
Balantidium coli 148
birth 8, 16, 22, 27, 39, 49, 66, 72, 82, 90, 100, 108, 119, 120, 126, 134, 144; *see also* labour; parturition
'bloat' 53, 103, 120, 122, 126, 129
Bordetella pertussis 138, 148
'burping' 11, 112, 137
bushbaby
   Senegal or lesser, *see Galago senegalensis*

Caesarian section 40, 109, 112, 120
Callitrichidae 3
Callithrix jacchus 3, **39–47**
candidiasis 45, 68, 122
Cebidae 3
Cebuella 53
Cercopithecidae 3
Cercopithecus aethiops 3, **81–7**
Cheirogalidae 3
chimpanzee, *see Pan troglodytes*
Clostridium tetani 138, 148
colic 11
colitis 54, 95
colobus monkey
   western black and white, *see Colobus polykomos*
colostrum 92, 110, 111, 112

constipation 11
Corynebacterium diphtheriae 138, 148
cross-fostering 12, 45, 55, 96, 115, 139, 148

Daubentonidae 3
dehydration 55, 77
diarrhoea 18, 22, 45, 83, 114, 122, 127, 135
diphtheria, *see Corynebacterium diphtheriae*
douroucouli, *see Aotus trivirgatus*
dystocia 77, 96, 144

enteritis 18, 44, 45, 95, 115, 148

fetal resorbtion 15, 77
folic acid 77

Galago senegalensis 3, **21–4**
   G. s. moholi 22
Galago crassicaudatus 22
gibbon, *see Hylobates lar*
gorilla 1
green monkey, *see Cercopithecus aethiops*
grivét, *see Cercopithecus aethiops*

Herpesvirus hominis 45, 55, 62, 68
Hylobates lar 3, **133–40**
   subspecies 133
hypothermia 11, 12, 77, 103

inbreeding 8, 90
incubator 11, 17, 43, 53, 55, 67, 75, 93, 100, 112–13, 128, 137, 146
Indriidae 3

labour 8, 40, 50, 72, 82, 90, 109, 144; *see also* birth; parturition
Lemur catta 9
Lemur fulvus 9
Lemur macaco 9
lemur
   lesser mouse, *see Microcebus murinus*
   roughed, *see Varecia variegata*
Leontopithecus rosalia 3, 44, **59–63**
   subspecies 59
Lorisidae 3

Macaca arctoides 3, **99–104**
Macaca mulatta 3, **89–98**, 99
macaque
   rhesus, *see Macaca mulatta*
   stump-tailed, *see Macaca arctoides*
marmoset
   common, *see Callithrix jacchus*
measles 55, 96, 115, 129, 148
meningitis 54, 148
metabolic rate 8, 16, 28, 42
metabolic weight 3
Microcebus murinus 3, **15–19**
   M. m. rufus 16
milk composition
   amino acids 73
   ash 9, 41, 51, 60, 73, 83, 92, 110, 145
   calcium 41, 110
   carbohydrate 9, 22, 41, 73, 83, 92, 100, 110
   casein 51, 110
   chloride 41, 110
   dry matter 9, 22, 41, 42, 51, 73, 83, 92, 100, 110, 145
   fat 9, 22, 41, 42, 60, 73, 83, 92, 100, 110, 111, 145
   fatty acids 41, 60, 73, 111
   lactose 9, 22, 41, 51, 60, 73, 110, 145

milk composition (*cont.*)
  lysozyme 110
  magnesium 41
  minerals 60, 73
  osmotic pressure 41
  phosphorus 41, 110
  potassium 41, 110
  protein 9, 17, 22, 28, 41, 42, 51, 60, 73, 83, 92, 100, 110, 145
  sodium 41, 110
  whey 51, 110
mumps 148

Nasalis larvatus 3, **119–23**

owl monkey, *see Aotus trivirgatus*
Old World monkeys, *see* Cercopithecidae

Pan troglodytes 3, **143–50**
  subspecies 143
Papio cynocephalus 3, **107–16**
  other *Papio* species 107, 108, 109, 110
  parturition 15, 18, 22, 23, 39, 49, 50, 67, 72, 82, 86, 90, 100, 108–9, 114, 119, 129, 134, 139; *see also* birth; labour
Perodicticus potto 3, **27–30**
pneumonia 18, 34, 42, 44, 45, 54, 68, 74, 77, 93, 95, 114, 136, 138, 146, 148
poliomyelitis 96, 138, 148
potto
  Bosman's, *see Perodicticus potto*

Pongidae 3
premature births 12, 68, 77, 78, 83, 92, 103, 109, 112, 113, 129, 134, 139
Prosimii 3
proboscis monkey, *see Nasalis larvatus*

rubella 148

Saimiri sciureus 3, **71–9**
  subspecies 71
Salmonella 12, 112; *see also* salmonellosis
salmonellosis 103
Saguinus species 3, 54, 55, 60
Saguinus oedipus 3, **49–57**
septicaemia 54
small pox 138
squirrel monkey, *see Saimiri sciureus*
stillbirth 12, 44, 54, 68, 72, 76–7, 85, 90, 95, 96, 102–3, 114, 126, 129, 148
supplementary feeding 23, 42, 45, 55, 103, 138
surrogate mother 11, 29, 43, 45, 52, 53, 56, 67, 75, 76, 113, 121, 128, 137, 138

tamarin
  cotton-top, *see Saguinus oedipus*
  golden lion, *see Leontopithecus rosalia*

tarsier
  Horsfield's or western, *see Tarsier bancanus*
Tarsier bancanus 3, **33–6**
  T. spectrum 33
Tarsiidae 3
temperature
  ambient 11, 17, 43, 53, 55, 62, 67, 75–6, 93, 102, 112–13, 114, 121, 128, 137, 146
tetanus, *see Clostridium tetani*
Theropithecus gelada 107
Toxoplasma 73
tuberculin test 12, 96, 138
tuberculosis 115, 138
typhoid 138

umbilicus 11, 113, 134, 144, 146

vaccination 55, 78, 96, 115, 129, 138, 148
Varecia variegata 3, **7–12**
  subspecies 7, 8, 10, 11
vervet monkey, *see Cercopithecus aethiops*
vitamin C 34, 73
vitamin D_3 16, 40, 45, 68, 73
vitamin E 66
vomiting 121, 122

whooping cough, *see Bordetella pertussis*

yellow fever 115